Taconite Dreams

Taconite Dreams

*The Struggle to Sustain Mining on
Minnesota's Iron Range, 1915–2000*

JEFFREY T. MANUEL

University of Minnesota Press
Minneapolis | London

A different version of chapter 1 was published as "Mr. Taconite: Edward W.
Davis and the Promotion of Low-Grade Iron Ore, 1915–1955," *Technology
and Culture* 54, no. 2 (April 2013): 317–45. Copyright 2013 The Johns Hopkins
University Press.

Published by the University of Minnesota Press
111 Third Avenue South, Suite 290
Minneapolis, MN 55401-2520
http://www.upress.umn.edu

A Cataloging-in-Publication record for this book is available from the
Library of Congress.
ISBN 978-0-8166-9429-7 (hc)
ISBN 978-0-8166-9430-3 (pb)

Printed in the United States of America on acid-free paper

The University of Minnesota is an equal-opportunity educator and
employer.

21 20 19 18 17 16 15   10 9 8 7 6 5 4 3 2 1

It is a privilege to thank the many people who contributed to this book. Although my name is on the cover, it could not have been written without the assistance of many supportive institutions, friends, and family members.

My research was generously assisted by funding from several institutions. Funding for my early research was provided by a Norman Johnston DeWitt Fellowship and a Doctoral Dissertation Fellowship at the University of Minnesota. A summer grant from the University of Minnesota's Graduate Research Partnership Program allowed for additional research trips to the Iron Range. A STEP Grant from the Office of Research and Projects at Southern Illinois University Edwardsville afforded me time to complete the manuscript.

My early research and writing benefited tremendously from the assistance of mentors and colleagues in the departments of history and American studies at the University of Minnesota. I am especially grateful to Lary May, who championed this research from the beginning and routinely steered me back to key questions about the Iron Range's culture. Elaine May also supported the project in its early stages and helped me to better situate the Iron Range's regional history within the broader stream of twentieth-century U.S. history. Tom Wolfe pushed me to think about the Iron Range in unconventional ways, and Tracey Deutsch helped to ground my research in the longer history of capitalism. Tom Misa shared his insights and deep knowledge of the history of technology. I also appreciated the insights and friendship of numerous colleagues in Minneapolis, especially Lisa Blee, Caley Horan, Danny LaChance, Ryan Murphy, Rachel Neiwert, Tim Smit, Jason Stahl, Brian Tochterman, and Andy Urban.

Friends and colleagues at Southern Illinois University Edwardsville supported later research and writing. Members of the history department writing group, especially Laura Fowler, Jessica Harris, Bryan Jack, Jennifer Miller, and Katie Sjursen, helped keep me on schedule while making revisions. Buddy Paulett read and commented on my book proposal even when he could have easily declined. Carole Frick was especially helpful in balancing research with the demands of teaching, and Jason Stacy always has an open door for chats about history or biking. I couldn't ask for a friendlier group of colleagues.

Several other scholars generously donated their time and expertise in the form of advice, comments, and helpful suggestions. I especially thank Steven High for encouraging me to study the history of deindustrialization at an early stage in my career. At various conferences over the years I received helpful suggestions and critiques from Bernie Carlson, John McNeill, Judith Stein, Frank Uekoetter, and George Vrtis. Finally, Josef Barton first gave me the idea to study Minnesota's Iron Range through the lens of deindustrialization.

I consulted numerous archives during my research for this book and a few archivists stand out for special thanks. At the Iron Range Research Center in Chisholm, Minnesota, Scott Kuzma and Philip Deloria were essential in digging up archival material on postwar Iron Range history. At Iron Range Resources, Sheryl Kochevar went far above and beyond her duties to make me feel at home as a researching historian. Iron Range Resources Commissioner Sandy Layman also gave me generous access to Iron Range Resources' archival materials. One of the great pleasures of writing about Minnesota's history is working with the Minnesota Historical Society in St. Paul, which offers a model for state historical societies to emulate. I especially thank Debbie Miller and Patrick Coleman for help in navigating Iron Range archival materials at the Minnesota Historical Society. Pieter Martin at the University of Minnesota Press has been a friendly and helpful guide for a first-time author. At the Minnesota Geological Society, Mark Jirsa stepped in to help with the map in the Introduction and offer a geologist's insight into the Iron Range's history.

Most important, I humbly thank the people who support and encourage me day in and day out. My parents, Kerry and Hilde Manuel, supported the research that went into this book in more ways than I can remember. On a deeper level, many of the arguments presented here were first developed during car trips or long political discussions with my parents. My in-laws, Rick and Diane LeBlanc, have been incredibly supportive over many years of research and writing. Susanne LeBlanc deserves special thanks: she has been a bedrock of love and support, a wonderful friend, and a check on my craziest ideas. A few words of acknowledgment are scant payback for her years of love and support. Finally, my daughter Nathalie arrived just when I was finishing the manuscript. She has given me the focus to complete this book and brought more joy to my life than I could have imagined. Thanks to all of you.

INTRODUCTION

An Industrial Mining Region in a Postindustrial Age

Nineteen fifty-nine was a good year to leave the Iron Range. In the fall of that year, before the region's notoriously cold winter set in, young Robert Zimmerman hit the road, headed south for Minneapolis and the University of Minnesota. He was leaving behind his family's modest home in the iron-ore mining town of Hibbing, Minnesota. The boxy clapboard home squatted in a neighborhood of modest homes. From the yard he could see the giant piles of overburden—small mountains of earth stripped to get at the valuable ore beneath—that ringed Hibbing. Zimmerman, later known to the world as Bob Dylan, would go on to fame and fortune as a singer and songwriter. But in the fall of 1959 he was one of many young people fleeing the iron-ore mining district. Hundreds of young men and women from Hibbing and the other iron-ore mining towns strung along the Mesabi Range in northeastern Minnesota made a similar trek, loading up their cars or catching a bus to head south. They left for Duluth, St. Paul, or Chicago. They left for jobs or college. Many of them would never permanently return.[1]

Nineteen fifty-nine was, for others, a good year to stay on the Iron Range. Ninety miles east of Hibbing, on the shore of Lake Superior's cold, glistening waters, the Reserve Mining Company's enormous new taconite mill was operational. The E. W. Davis Works used highly advanced milling technology to produce manufactured iron-ore pellets for the steel industry. The mill was a brilliant feat of mineral engineering. Piano-sized boulders of hard taconite rock entered the mill and, after precise crushing, grinding, and agglomeration, they emerged as marble-sized pellets manufactured to the specifications of the steel mills that used them in their blast furnaces as raw material for steelmaking. It had taken decades of engineering to perfect this system for using low-grade iron ore. By 1959, the mill had been open for four years and employed thousands of workers in good-paying jobs. The mill was only the first of many new low-grade iron-ore mines and mills on the Iron Range. The business press breathlessly anticipated a billion dollars of new investment in iron ore. For Iron Range residents, the new technology meant jobs and, most important, a future in the region they called home.[2]

Was it better to leave or to stay? This question ran through the politics, economy, and culture of the Lake Superior iron-ore mining district in the second half of the twentieth century. It was discussed in economic development meetings, political debates, and at countless family dinners. Each path had its risks. Those who left knew they would be isolated from family and friends, making more money, perhaps, but at the cost of separation and a sense of being uprooted. Staying was also risky. Sticking around meant placing a bet that the region's economic engine, the iron-ore mines, would continue into the future. Even for those who didn't work in the mines, mining employment drove related service-sector jobs, paid for government services, and held the region's culture together as a coherent whole.

Beyond the Iron Range, the tension between leaving or staying is a useful way of imagining the challenges facing industrial regions in the United States in the second half of the twentieth century. Were the nation's industrial regions—from Pennsylvania's anthracite mines to Detroit's auto-manufacturing plants to the Iron Range—vestiges of an older era that needed to be transformed or even abandoned in a brave new future of postindustrial capitalism? Or were they crucial bedrock for the nation—the places where things were made—that needed to be supported with financial, political, and cultural assistance? More broadly, regions such as the Iron Range were torn between the inevitability of decline in the mining industry, and in all industry, and the desire to keep an industry going into the future, to overcome all natural limits. Industrial-scale mining is an especially poignant example of this tension, but it haunts all industrial regions in the postindustrial age. As historians Jefferson Cowie and Joseph Heathcott point out, during the twentieth century heavy industry fooled many people into thinking it was permanent because the giant machines and open-pit mines appeared fixed and immovable. Gazing across man-made chasms and hundreds of tons of heavy equipment, people convinced themselves that industry was a system that could never break down. And yet it did.[3]

This book describes a century of efforts to prevent decline and deindustrialization on Minnesota's Iron Range. Working in industrial research laboratories, the state legislature, economic development offices, and even history museums, Iron Range residents waged a war against obsolescence in the twentieth century. In this battle they were caught between two conflicting beliefs about themselves and the region they called home. On the one hand, many were certain that the Iron Range, the steel industry it was part of, and the national industrial economy they contributed to were essential to the modern United States. It was impossible to imagine a modernizing America, they thought, standing astride the globe by the middle of the twentieth century, without their essential backbone of industrial labor, raw steel, and the tough towns where things were made. This vision imagined a vital, modern nation built on the labor and production of a patriotic working class.[4]

On the other hand, many residents acknowledged that the economics of the global steel industry did not bode well for their region. Many understood the realities of new foreign ore fields that were twice as rich as theirs worked by miners paid half as much. They also saw a political culture moving quickly to distance itself from the gritty blue-collar steel and mining towns. Thus, Iron Range residents faced the hard realities of globalization and marginalization throughout the late twentieth century. The tension between these two magnetic poles—one honoring blue-collar communities as central to the modern United States, and the other offering an increasingly small range of depressing options in the face of global competition and postindustrial culture—is crucial to understanding the history of the modern Iron Range and America's industrial heartland.[5]

The Iron Range's history in the second half of the twentieth century was driven by the challenges of industrial decline and residents' stubborn efforts to fight against it using all available tools. Widely considered the globe's richest and most productive iron-ore mining district at the beginning of the century, the Iron Range struggled with the challenges of mineral depletion, rising global competition, population loss, and cultural displacement throughout the twentieth century. At no point, though, did residents give up and accept decline. The fight against it took many forms over the decades, but it was animated by a stubborn refusal to accept that the region would devolve into a string of mining ghost towns.

Although the Iron Range's experiences were unique to its iron-ore mining industry and history of European immigration and political radicalism, understanding the costs and consequences of its fight against decline provides greater insight into questions of central importance in modern American history. How did industrial regions respond to long-term economic decline in the twentieth century? What options were available in the fight against decline and how did politicians and residents decide on the eventual course taken? What were the consequences of fighting back against deindustrialization? Who gained and who lost in trying to keep an industrial economy afloat in the face of decline and globalization?

The Iron Range Region

Minnesota's "iron deposits," historian Theodore Blegen wrote, "go back into the recesses of geologic time." Geologists theorize that three major geological processes created the iron-ore deposits of northeastern Minnesota. First, as mountains eroded during the Lower Proterozoic time, iron flowed into a vast inland sea that covered today's northeastern Minnesota. Oxygen, likely from marine algae, combined with the iron to form the rich iron-ore deposits. Next, a period of mountain building emptied the inland sea and pushed the deposits skyward. The iron deposits were folded and contorted as the earth buckled. As the mountains eroded, some of the deposits were leached by the wind and water to leave exceptionally rich concentrations of iron. These pockets of rich ore become the high-grade iron-ore mines that were first exploited on the Iron

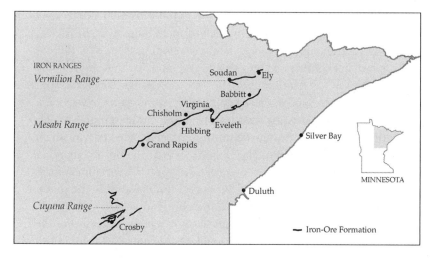

Minnesota's Iron Range region. Base map courtesy of Minnesota Geological Survey. Designed by Susanne LeBlanc.

Range. Finally, retreating glaciers left earth and debris on top of the iron formation. This skin of earth was the overburden that was later removed by miners to reach the rich ore underneath.[6]

The Iron Range region's modern history began in the late nineteenth century. The area that would later become the Iron Range was largely uninhabited as late as the early nineteenth century. Ojibwe people hunted in the forests of what would become northeastern Minnesota prior to the arrival of American industry. Lumberjacks arrived in the second half of the nineteenth century to cut the region's white pine forests. Northeastern Minnesota's vast pine forests provided raw materials for Minneapolis's and Duluth's burgeoning sawmills. The region has a long history of resource-extractive industry.[7]

The modern Iron Range took shape after the discovery of valuable metals in the region, first gold and then iron ore. Prospectors arrived in northeastern Minnesota in large numbers in the 1860s following reports of gold on the shores of Lake Vermilion. Once there, they noticed large iron deposits but found little gold. Most gold prospectors abandoned the region, leaving a handful of miners hoping to exploit the region's iron-ore deposits. By the 1870s, the expanding and maturing iron and steel industry needed far larger quantities of iron ore and the Iron Range's known deposits became far more valuable. An expedition led by former gold miners in 1875 mapped the main ore deposits of the Vermilion Range. They quickly enlisted the help of financiers, including the wealthy easterner Charlemagne Tower, to begin developing the area's iron-ore deposits for commercial mining and shipment. By 1884, mines were operating at seven locations on the Vermilion Range, which was connected to Lake Superior via a new railroad. Mining on the Vermilion Range involved blasting hard rock iron ore in underground mines, a process similar to iron-ore mining methods used previously in Michigan's Upper Peninsula.[8]

It was only after several years of mining on the Vermilion Range that explorers discovered the far larger Mesabi Range to the south. Miners familiar with hard rock iron-ore deposits in other parts of the country initially dismissed the soft granular red dirt of the Mesabi Range because they assumed it could not be iron ore. When Henry C. Frick, chairman of Carnegie Steel, sent a trusted expert to the Mesabi Range in the 1890s, the expert reported that the area's dirt could not be iron ore. Nonetheless, a small group of Duluth businessmen, notably the Merritt family, moved quickly to take ownership of large tracts of land on the Mesabi Range at the end of the century. This land proved to be not only iron ore, but an especially rich and easily accessible vein of ore.[9]

After the Merritts developed the initial mines of the Mesabi Range and built a railroad to bring the ore to Lake Superior, their business interests were damaged by the Panic of 1893. The small independent mines were subsequently consolidated into a few large corporations working with steel companies. By the late 1890s, the mines of the Mesabi Range were held by the corporate titans of the industrial era, including John D. Rockefeller, Andrew Carnegie, James J. Hill, and Elbert H. Gary. When J. P. Morgan consolidated Carnegie Steel into the massive U.S. Steel in 1901, Rockefeller sold his Iron Range holdings to the new company, making the Iron Range the major supplier of iron ore for the U.S. steel industry in the early twentieth century. Thus, the Iron Range has been an essential part of the U.S. steel industry and offers an excellent vantage point to examine the industry's fate throughout the twentieth century.[10]

During the first half of the twentieth century, Minnesota's Iron Range boomed as the region became the primary supplier of iron ore for the U.S. steel industry. The Mesabi Range's massive open-pit mines dwarfed other iron-ore mining regions, both in the amount of ore they produced and the sheer size of the open pits. Towns and then cities grew alongside the open-pit mines as workers from

By the first decades of the twentieth century, the Mesabi Range's open-pit iron-ore mines, such as the Mahoning Mine in Hibbing pictured here circa 1920, were among the largest and most technologically advanced mines in the world. Minnesota Historical Society Collection, HD3.112 r75.

around the globe arrived in the region. More than a hundred thousand residents made the Iron Range their home by 1920. Tax money from the rich iron-ore mines built ornate public buildings and provided for public welfare across the Iron Range. The iron-ore mines were harmed by the Great Depression of the 1930s. Most of the mines shut down and unemployment rates reached 70 percent in some parts of the Mesabi Range. Yet the demand for war materials in the 1940s led to a renewed boom in the region's mines. The Iron Range again flourished during World War II as the region's mines furnished raw materials for the steel-clad armed forces. By the dawn of the postwar era, the region had already experienced several cycles of boom and bust.[11]

The steel industry was at the heart of the U.S. economy in the twentieth century. Steel was perceived to be so central to the United States during the years after World War II that business leaders, labor officials, and economic policy makers often used the steel industry's production as a barometer for the national economy, a policy that historian Judith Stein calls "steel fundamentalism." Because steel was the basic building block of many of the postwar

Mesabi Range mines operated at full capacity during the 1940s to meet wartime demand for iron ore. In this photograph, miner John Palumbo Jr. stands on the edge of Hull-Rust-Mahoning mine. Iron Range miners unionized under the United Steelworkers during World War II. Photograph by John Vachon, August 1941. Library of Congress, Prints and Photographs Division, FSA/OWI Collection, LC-DIG-fsa-8c20455.

world's consumer goods, from automobiles to refrigerators to children's toys, and was needed for productive industries such as construction, government officials believed that by working with the steel industry they could shape the nation's macroeconomic fortunes. Through its connection to the steel industry, the Iron Range was important to postwar debates about America's overall industrial policy and the relationship between government and heavy industry.[12]

Immigration, especially from Eastern Europe, also defined the Iron Range's history. Iron-ore mining in the Mesabi Range's open pits was labor-intensive work and immigrant laborers from halfway around the world flocked to the Iron Range in search of jobs. Once there, they created a distinctive culture of immigrant communities that the Iron Range has retained into the twenty-first century. Celebrations of ethnic identity, including traditional costumes, music, and food, are common on the Iron Range. Yet most immigration to the region ceased by the mid-twentieth century. Because of declining job opportunities in the iron-ore mines, immigrants who came to the United States after 1965 largely bypassed the Iron Range.[13]

A century of mining, immigration, and hard times contributed to a distinct Iron Range identity and culture. "Rangers," as locals call themselves, make up a distinct subculture within Minnesota's demographic landscape. Rangers celebrate their multigenerational legacy of hard work in the mines, the harshness of the region's climate and landscape, and its melting-pot ethnic culture. Historian and Iron Range resident Marvin Lamppa defines the Iron Range identity as "a sense of place and shared identity . . . among the working people of the Range" that included "pride in one's ethnic heritage" and knowing that "their roots run deeper than the ores." Iron Range journalist Aaron Brown describes his Iron Range identity as "one part geography and one part attitude." It is a contradictory identity, Brown notes, where "on one side you've got pigheadedness, decline, and dependency on outside forces to change things. On the other you have hard work, survival, and innovation." Iron Range identity is often strong enough that residents hold on to it even after emigrating out of northeastern Minnesota. It is common to find Iron Range clubs and get-togethers in the Twin Cities and other regions where Iron Range residents moved for work. In the twenty-first century, social media sites offered scattered Iron

Rangers a place to convene and reaffirm their identity. The Iron Range stamps its residents with a strong regional identity that often is not wiped away by distance from the region.[14]

Many attributes of this Iron Range identity would be familiar in other regions with strong traditions of working-class ethnic solidarity or mining work. Characteristics such as stubborn tenacity in the face of hard times, labor liberalism, or regional insularity can be found throughout other mining regions in the United States, including Appalachia, Pennsylvania's coal country, or Michigan's Upper Peninsula. Iron Range identity and culture was unique within Minnesota, however, which made the region stand out from other rural parts of the state in the twentieth century. Differences between Rangers and the rest of Minnesota occasionally put the region at odds with state politicians in St. Paul as well.[15]

Defining the precise boundaries of the Iron Range has always been difficult. Early in the twentieth century, observers spoke of the Lake Superior region's iron ranges in the plural, emphasizing mining activity in several areas of Minnesota, Wisconsin, Michigan, and Ontario. Mining activity and local identity were tied to a specific range, such as the Vermilion, Cuyuna, or Mesabi. Historian Marvin Lamppa argues that the term "Iron Country" best describes all of northeast Minnesota because iron-ore mining was so influential for the region's early development. Yet the singular term "Iron Range" was widely used by the end of the twentieth century to connote those parts of northeast Minnesota with an active mining economy. This was partly the result of declines in mining on the smaller Vermilion and Cuyuna ranges. In truth, this singular Iron Range region was the Mesabi Range and the two terms came to be used interchangeably by the end of the twentieth century. Many casual observers forgot about past mining activity on the Vermilion or Cuyuna ranges. As described in chapter 5, clarifying and simplifying the Iron Range's definition and boundaries was the result of tourism and heritage promotion in the late twentieth century. To be understandable—and visitable—to outsiders, the Iron Range had to first be clearly defined. Defining which parts of Minnesota are and are not part of the Iron Range has taken up a good deal of time and energy over the years, but the crucial point is that the region's boundaries shifted over time as various actors—miners, politicians, resort owners, and even scholars—redefined the Iron Range to suit their own purposes.[16]

Technology and the Environment

Three main themes run through the Iron Range's history in the second half of the twentieth century. The first is a twinned history of technology and the environment. Technological innovation was the foremost way in which Iron Range residents and politicians attempted to fight off industrial decline. Like so many Americans in the twentieth century, they turned to new technologies to solve their problems. Specifically, the Iron Range turned to a new technology for mining and milling low-grade taconite, the previously worthless rock that was abundant throughout the region. Taconite saved the Iron Range from depleted hematite mines and new global competition. Beginning in 1913, Edward W. Davis, a professor at the University of Minnesota's School of Mines, began the long process of turning worthless taconite into valuable iron ore. His work came to fruition decades later when several major steel firms invested millions of dollars to build taconite mines and mills on the Iron Range in the 1950s and 1960s. Chapter 1 describes the decades of work required to perfect low-grade taconite-ore milling and shift the region's iron-ore industry from older hematite mines to the new taconite mines. From the perspective of the 1960s, technology appeared to save the Iron Range. Taconite promised decades—centuries, even—of future iron-ore production in the region. And best of all, taconite technology seemed to be a painless fix, requiring few of the hard choices that typically accompany declining industries.

Yet, a longer historical lens reveals that taconite technology was hardly painless. Implementation of the new technology, as well as the overall shift in the region from older high-grade hematite ore mining to new low-grade taconite-ore mining, had numerous unintended consequences. Older mines could no longer compete with the new low-grade ores and these mines quickly went out of business, throwing their workers out as well. The region's geography was altered, too, because the taconite mines were located away from traditional population sources.[17]

The most significant unintended consequence that sprang from taconite technology was environmental pollution. Taconite mines and mills were models of efficient high-throughput mining. Taconite was a form of nonselective mining, which took huge amounts of rock from the earth and processed it in enormous mills to create valuable iron ore. Yet the mills' efficiency in turning low-grade ore into iron made them similarly efficient at producing enormous quantities of tailings waste and air pollution. Problems with taconite's environmental impacts crystallized in the late 1960s with a lawsuit against the Reserve Mining Company's E. W. Davis Works. This taconite mill, perched on the edge of Lake Superior in Silver Bay, Minnesota, was dumping sixty-seven thousand tons of taconite tailings into the lake each day by the 1960s. The plant became the focus of a bruising, decade-long pollution lawsuit. Judges alternately closed and reopened the plant before Reserve Mining was forced to switch to on-land tailings deposit in the 1980s. Although much has been written about this famous trial, chapter 3 argues that it must be understood within the context of taconite's development as a technological fix for decline on the Iron Range. Taconite's seemingly miraculous technological and engineering feats were possible only because they shifted the costs onto the environment. When those costs could no longer be ignored—as when spreading taconite tailings began discoloring Lake Superior—people reassessed the overall costs and consequences of industrial development. Chapter 3 also describes the trial's profound effect on local residents. For many miners and their families, the case was their first glimpse of the new environmental movement that would challenge their livelihood. Environmentalism remains deeply contested on the Iron Range into the twenty-first century.

Historians of deindustrialization and industrial decline have not yet devoted significant attention to efforts to find a technological fix for decline. Existing histories of declining industrial regions in the United States emphasize the personal challenges faced by workers and their families, the economic costs for the region, and the political efforts to stymie decline. All of these efforts occurred on the Iron Range, yet they were secondary to the attempts to develop a technological fix that would resolve the challenges of decline without forcing hard choices.[18]

Politics and Policy

The second theme running through the Iron Range's recent history is the role of politics and policy as tools for fighting industrial decline. Economic development to prevent decline consumed Iron Range politics in the second half of the twentieth century. From the local to the national level, politicians attempted to fix the region's struggling economy by tweaking the tax code, offering direct sub-

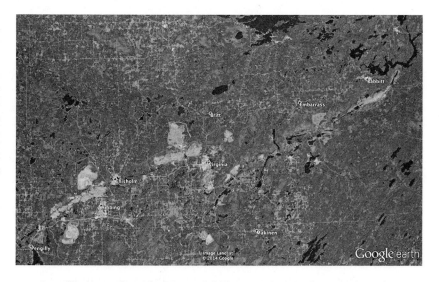

A satellite image shows the extent to which taconite mining has transformed the Iron Range landscape. The large light-colored patches are open-pit mines or holding ponds for the tailings produced in large quantities by the taconite process. Courtesy of Google Earth.

sidies to businesses, and creating tourist sites such as museums. Elections and speeches were the most visible examples of political approaches to fighting decline, but more significant efforts occurred in the less well-known fields of tax and economic development policy.

Taxation has long been a central lever for public policy in the United States. At the national level, federal taxation was both a reflection of the nation's priorities and a steering wheel for politicians trying to direct future development.[19] Tax policy was also an important tool for shaping the Iron Range. The iron-ore mining industry was deeply affected by changing taxes on mining. Taxes on mining were used by state and local officials as a tool to ameliorate the worst excesses of an industrialized resource-extractive industry. The mining companies also fought vigorously against increased taxation, arguing that their investment in the region was dependent on low taxation. Debates over taxation on the Iron Range thus served as proxies for broader debates about the region's economic future and what, if anything, the iron-ore mines owed to the cities and state where they operated.

Chapter 2 describes one crucial debate over mining taxation:

the mid-1960s dispute over adding a so-called taconite amendment to Minnesota's constitution. Several mining firms had invested in taconite by the early 1960s, but other companies insisted that high taxes on taconite and Minnesota's history of raising taxes on iron ore prevented them from building new mines and mills. In response, taconite's supporters, led by Democratic-Farmer-Labor (DFL) politicians from the Iron Range, mounted a statewide drive to amend the constitution. The amendment prevented taconite taxes from rising disproportionately to other business taxes. The taconite amendment revealed a split between committed liberals, who were dead set against preemptively cutting taxes for some of the world's largest and most profitable corporations, and labor union members, who took a pragmatic approach that emphasized jobs. The amendment dispute cleaved Minnesota's DFL Party throughout the first half of the 1960s. Eventually, the forces in favor of the amendment won, but it set the stage for future arguments against corporate taxation that continue today.

Like many American industrial regions during the decades after World War II, the Iron Range turned to economic development policy in response to economic decline. By promoting economic diversification, luring new industries to a region, or financing innovative or entrepreneurial ventures, economic development policy offered a pragmatic policy tool kit for regions like the Iron Range. Unlike many other regions, however, the Iron Range received strong institutional support for its economic development projects through an active and well-funded state agency, the Iron Range Resources and Rehabilitation Board (IRRRB). As described in chapter 4, the IRRRB was formed in the early 1940s with a mission to diversify the Iron Range's economy and ensure that tax money from mining was spent on long-term investments. Yet the IRRRB was enmeshed in controversy and uncertainty from its birth. The board spent its first years attempting far-flung development projects, few of which paid off. Then, in the late 1950s and 1960s, it became a model for a short-lived national agency designed to support declining regions: the Area Redevelopment Administration (ARA). But the ARA was soon mired in controversy and abandoned. This brief federal program was a missed opportunity to deal with declining regions, such as the Iron Range and Appalachia, on a national level. With economic development policy left to the regional and local levels, the IRRRB struggled to develop new industries on the Iron Range, often coming under fire for well-publicized boondoggles such as its support for a Hibbing chopsticks factory in the 1980s. Yet economic development has been the predominant approach to and framework for responding to industrial decline on the Iron Range. The IRRRB was not unique in this regard. Throughout the postwar era, economic development agencies proliferated within municipal, regional, and state governments and economic development became a central organizing principle of governmental action in many declining regions.

Yet, relying on regional development agencies such as the IRRRB to coordinate the response to deindustrialization had significant consequences. This local focus foreclosed national or even global responses to industrial decline and ultimately limited the horizon of possibilities to a narrow band of local and regional options. The local emphasis of economic development policy also inadvertently reinforced the problematic idea that industrial areas such as the Iron Range were independent and self-sustaining working-class communities, connected to national or global processes only to the extent of their victimization. The local emphasis of economic development policy also illustrates the noticeable lack of a national policy in response to deindustrialization in the postwar United States. By relying on state and local agencies such as the IRRRB, planners and bureaucrats at the federal level during the postwar era left economic development—and thus the response to deindustrialization—off of the national agenda. The problems facing declining regions never became a sustained federal issue in the second half of the twentieth century, as they were during the New Deal era, and states and regions were left to pursue economic development projects as best they could according to their available resources.[20]

Beneath the details of taxation and economic development policy, the Iron Range's modern history also suggests that the connections forged between the labor movement and liberalism on the Iron Range in the twentieth century were surprisingly brittle. In few places in the United States were labor and liberalism as tightly connected as they were on the Iron Range, where residents vehemently argued that their iron ore, their labor, and their votes were at the heart of the nation's political and economic system. Unions, especially the United Steelworkers of America, the union that represented iron-ore miners, joined with liberal politicians in the state's Democratic-Farmer-Labor Party to create a political juggernaut on the Iron Range. More than any other development, however, deindustrialization and long-term industrial decline hollowed out older connections between industrial labor and liberalism. The result of this process was a postwar political culture that made industrial labor increasingly marginal as a vital contributor to the region's and the nation's future. By the twenty-first century, liberal politicians still made it a point to shake hands with union leaders, but they now met in front of closed factory gates.[21]

Even in a region deeply committed to maintaining the connection between liberal politics and industrial labor, long-term industrial decline corroded the strong ties between labor and liberalism. Challenges came from multiple directions. The miners and their families on the Iron Range proved to be far more pragmatic in their politics than many liberal politicians expected. When offered a choice between reining in corporate power or protecting their paychecks or job security, the miners consistently chose the latter. In turn, liberal politicians, at both the federal and state levels, had surprisingly few answers to deindustrialization. In the absence of tangible and widespread economic benefits for supporting liberal policies and politicians, the labor-infused liberalism that once permeated the Iron Range was soon hollowed out to a shell of itself.

Memory

The third major theme threaded through recent Iron Range history is memory—the contested memory of industrial mining on the Iron Range as the region struggled to define itself in the late twentieth century. Was the Iron Range moving away from mining and toward a postindustrial, service-sector economy? And, if so, how should residents remember the iron-ore mining industry that was the region's lifeblood in past decades? Was mining an obsolete relic to be overcome in the future? Or was mining a modern, vital industry that needed support so it could grow into the future?

Debates over the region's future often played out as struggles to understand the region's past. Historians have noted how memory and history are powerful venues for conflict in American life. In the aftermath of the Civil War, for instance, Americans debated the meaning and memory of the war, ultimately enshrining a memory of the war as a tragic sacrifice of white men and erasing memories of the war that emphasized its radical emancipationist vision. Memory of the past has proven controversial in industrial regions as well. Memorials and museums have been used to revitalize older industrial cities (often with lackluster results) and the turn toward industrial heritage has raised thorny questions about whether industry is a product of the past or a concern for the future.[22]

Historians, especially public history professionals working in museums and archives, played a crucial role in depoliticizing industrial decline during the last decades of the twentieth century. Across the Iron Range, perceptions of the iron-ore mining industry gradually coalesced around a sense that the region's mining industry was somehow nostalgic by the end of the century. This did not happen by accident. In creating museums, collecting archives, and writing historical narratives, historians and heritage professionals wrapped the Iron Range's mining industry in a patina of industrial heritage. As a relic of nostalgic heritage, iron-ore mining could be simultaneously honored in the past and ignored when planning a postindustrial future for the region. History and nostalgia proved especially effective in overcoming the deep hostility and unease felt by many miners toward development plans that did not promote a continued industrial economy on the Iron Range. By promising to celebrate iron-ore mining's rich history in the region, heritage professionals and historians soothed industrial workers' concerns about the future by honoring the past.

Chapter 5 describes the most significant battlegrounds over memory on the postwar Iron Range. Amid conflict over the future direction of the Iron Range's economy, which pitted mining supporters against those promoting a postindustrial service and tourist economy, the region turned to heritage tourism as a compromise solution. The centerpiece of heritage tourism on the Iron Range was a museum and entertainment complex originally called the Iron Range Interpretative Center (today called Minnesota Discovery Center but generally known by its 1980s name, Ironworld). Ironworld was intended to attract tourists who would admire the history of industrial mining presented in the museum. Yet the museum was immediately bogged down in controversies concerning its presentation of history, its low attendance, and its focus. Asked to serve two masters, history and tourism, it satisfied neither. The museum was a microcosm of the larger problems of historical memory on the Iron Range and in other declining industrial regions, where residents were torn between a desire to honor the industrial past and the challenge of moving into a postindustrial future.

Deindustrialization on the Iron Range

Economic change on the Iron Range is usually described using a language of boom-and-bust business cycles typical of resource-extraction industries and one-industry communities. "Mining has always been a boom and bust industry on the Iron Range," notes a 2013 news report.[23] Others use metaphors of a roller coaster or turbulence to capture the region's volatile economy. While the language of booms and busts helps to explain the volatility of the Iron Range's mining economy, boom-and-bust rhetoric is problematic for several reasons. First, the term hides the fact that overall mining employment on the Iron Range declined steadily throughout the twentieth century. In 1944, more than eleven thousand workers labored at ninety-five iron-ore mines in Minnesota. By 2011, only seven mines were left in the state and they employed fewer than four thousand workers.[24] Iron-ore mining jobs were lost owing to mine shutdowns, depletion, foreign competition, and technological automation. But the end result was the same: fewer and fewer jobs in the region's mines and mills. Although there were significant cyclical upturns in the region's economy—a boom in employment and construction jobs during the 1970s, for instance, as new taconite plants were built—the long-term employment trend has been consistently downward. Boom-and-bust rhetoric obscures a long-term erosion of mining employment and offers a false promise that the region is just one boom away from the good old days. This is especially problematic for workers left reeling from permanent mine shutdowns where boom-and-bust talk holds out false hope that the mines will reopen. Second, boom-and-bust rhetoric naturalizes economic change in the iron and steel industry by ignoring the crucial role that politics and policy play in determining supply and demand in the industry. The iron and steel industry was considered central to national economic power and military might during the twentieth century. Supply and demand in the industry were manipulated by national and international policy, a fact that is obscured by boom-and-bust rhetoric naturalizing the ups and downs of the iron and steel markets. Finally, boom-and-bust rhetoric was (and is) used in problematic ways by those wishing to justify new industrial projects on the Iron Range. Supporters of new mining ventures hold out the promise of a new economic boom to drum up support for their agendas, knowing that there will be little accountability if the project later fails to live up to the original predictions. Taconite's supporters did this during the middle of the twentieth century and there is evidence that supporters of new forms of mining are doing this in the early twenty-first century, as discussed in the conclusion. Touting the potential economic benefits of new industry is not inherently problematic, but such predictions, which often overstate the economic benefits of new industry and underestimate the costs, build on a boom-and-bust psychology to justify and naturalize new mining ventures. Such rhetoric can easily shortcut evenhanded analysis and discussion of the costs and benefits of new mining ventures, a process that has happened several times in the Iron Range's history.[25]

In contrast to the rhetoric of boom and bust, this book argues that the Iron Range's modern history should be understood through the lens of deindustrialization, a term referring to the economic, social, and cultural changes resulting from the closure of heavy industry such as mining and manufacturing. It is impossible to fully understand the modern Iron Range without considering how deindustrialization shaped the region over the past half century.

The term "deindustrialization" came into widespread usage in the 1970s and has since been used by scholars to compellingly describe the economic, political, and cultural changes that followed the shutdown of heavy industry in regions around the globe. Deindustrialization was originally used in the aftermath of World War II to describe a (never implemented) plan to strip Germany of its industrial capacity so that the nation could not wage war again. There was little discussion of deindustrialization during the economic boom of the mid-twentieth century, despite the fact that many industrial regions such as the Iron Range and the country's inner cities were already feeling the pinch of factory closures during these years. Local deindustrialization in the early postwar era was covered up by optimistic rhetoric about national economic growth.[26] Discussion of deindustrialization roared back onto the scene in the late 1970s and early 1980s, however, as waves of factory shutdowns in North America and Europe flashed across the headlines. Initial scholarship on deindustrialization was infused with political cries to save the factories and working-class communities from abandonment and decline. Most of these efforts to prevent deindustrialization faltered. Especially in the United States, local fights against mine and mill closures attracted headlines and, occasionally, political support, but there were precious few success stories of communities that preserved local industries. Deindustrialization scholarship since the 1980s has documented the long-term consequences of mine and factory shutdowns, demonstrating how closures tore apart the social and cultural fabric of working-class communities in obvious and subtle ways. Deindustrialization thus offers a compelling interdisciplinary lens for analyzing the timing, causes, and consequences of industrial decline.[27]

Deindustrialization is also a useful way of understanding capitalism's deeply uneven regional effects. Several decades of scholarship reinforce the conclusion that the effects of capitalist economic change must be understood by looking at specific locales, such as neighborhoods, regions, or states, rather than entire national economies. A national focus—typically measured by broadbrush statistics such as gross domestic product—obscures regional changes that tell a very different story from national narratives. For instance, Thomas Dublin and Walter Licht demonstrate how deindustrialization and economic decline tore apart the economic and social foundations of Pennsylvania's anthracite coal-mining region during the middle of the twentieth century, an era typically associated with the booming national economy. Subnational industrial regions outside the United States have also been affected by deindustrialization. For instance, historian Christopher H. Johnson describes how southern France's Languedoc region suffered through deindustrialization in the late nineteenth and early twentieth centuries. Similar tales of regional deindustrialization have been documented across Western Europe and Great Britain. Minnesota's Iron Range should be added to the tragically long list of industrial regions in North America and Europe that have struggled with economic decline and the transition away from dependence on a single heavy industry.[28]

The Iron Range's twentieth-century history also contributes new insights into the broader history of deindustrialization in the United States and Europe. Scholars from multiple disciplines have written evocatively about deindustrialization in the twentieth century, but much of the existing scholarship focuses on the cultural or economic consequences of industrial job losses. The Iron Range's fight against deindustrialization in the twentieth century instead highlights the active role of politics and policy in fighting back against industrial decline. On the Iron Range, politicians at the local, regional, and state level clearly saw the looming problem of deindustrialization and took steps to prevent the region from declining into obsolescence. Such policies included active support for new mining technologies at midcentury as well as regional economic development policy channeled through the IRRRB. Too often, industrial communities such as the Iron Range are portrayed as innocent or naive victims of economic dislocation. In truth, Iron Range residents and their political representatives were well aware of the problems facing the region's mining economy and took active steps to ameliorate the worst consequences of deindustrialization.[29] Deindustrialization is often imagined as an all-or-nothing outcome: either the factory stays open and the surrounding community thrives, or the mill closes and collapse ensues. The truth on the Iron Range, and elsewhere in industrial America, was far more complex. Many mines closed and thousands of miners lost jobs in the twentieth century. But new technologies, active politicians, and gritty local residents hung on to the region's mining economy, for better and for worse.

ONE

A Technological Fix?

Edward W. Davis and the Creation of Taconite

On September 13, 1956, Edward W. Davis joined approximately six hundred other engineers, business executives, and local politicians to formally inaugurate the new Reserve Mining Company plant for milling low-grade iron ore. The guests gathered in one corner of the cavernous plant and sat on folding chairs in front of a temporary stage. At the podium, a speaker read a congratulatory letter from President Dwight Eisenhower, who wrote that Reserve Mining's plant would contribute to "the progress and security of the United States." Eisenhower's interest in a new mining facility hinted at the plant's larger significance to the Iron Range mining region of northeastern Minnesota and the national economy. The Reserve Mining Company's new facility was the opening wave of a wholesale change in the globe's iron-ore industry. More important for the Iron Range, the plant promised decades of continued industry and employment for the mining region.[1]

1

The ceremony dedicating Reserve Mining Company's new taconite mill at Silver Bay, Minnesota, in 1956. Edward W. Davis, the engineer who worked for years to perfect the taconite process, sits near the podium. The mill symbolized the opening of the taconite era on the Iron Range. Minnesota Historical Society Collection, HD3.27 p26.

The Reserve Mining Company mined and milled a low-grade iron ore named taconite, the name given to an extremely hard rock found in abundance throughout the Mesabi Range. As it exists geologically, taconite contains approximately 20 to 30 percent iron as magnetite, a type of magnetic iron. This was much less iron than hematite, a mineral found in other parts of the Mesabi Range that contained 50 to 70 percent iron. Rich hematite ores could be scooped out of the ground and shipped directly to blast furnaces as a raw material for steel. Low-grade or lean ores such as taconite, on the other hand, required extensive processing before they could be used by the steel industry. The rock had to be mined and then crushed to the consistency of flour. Using powerful magnets, the magnetic iron was separated from the surrounding non-iron rock, or gangue. Concentrated iron was then rolled into small pellets, a process known as agglomeration, which were fired to make them hard for transport.

Low-grade iron ores such as taconite revolutionized the global iron and steel industry in the second half of the twentieth century. The Reserve Mining Company was thus a bridge between two distinct eras in the history of iron-ore mining. The first era, which lasted until the mid-twentieth century, treated iron ore as a mineral resource that was mined where it occurred in the Earth's crust. This first era was dominated by fears of resource depletion and a frantic search for new ore fields. The second era, which continues today, relies on low-grade iron-ore deposits that can, through extensive industrial processing, be manufactured into high-iron products. Worries about depletion have been resolved in this era because low-grade ores are plentiful around the globe. But this abundance has come at a cost, specifically a tension between development of new mines and concerns about the ecological consequences of low-grade mining on a mass scale. The unintended consequences of taconite mining were not felt immediately, though. When the Reserve Mining Company opened its plant in 1955, the company and residents of the Iron Range were hugely optimistic for the future of taconite.

Reserve Mining's mill in Silver Bay, Minnesota, was named the E. W. Davis Works to honor the engineer who had done more than anyone else to create a taconite mining industry. Edward W. Davis was born in 1888 in Cambridge City, Indiana. He received a master's degree in electrical engineering from Purdue University and briefly worked on "magnetic phenomena" for Westinghouse and General Electric. In 1911, Davis moved to Minneapolis to take a new position as a mathematics instructor in the University of Minnesota's School of Mines. As this chapter describes, Davis used his institutional home at the University of Minnesota's School of Mines to dramatically change the iron-ore mining industry.

The rise of the taconite industry was an especially significant development for the Iron Range. Taconite was the first of several attempts to save industrial iron-ore mining in the region. Instead of facing decline, depopulation, and deindustrialization, taconite offered a grand technological fix for the Iron Range. It was a technological innovation that promised to obliterate any limits to nature or geology and allow mining in the region to continue indefinitely. On the Iron Range and in other declining industrial regions in the

Edward W. Davis in 1940. Minnesota Historical Society Collection, por 5957 p2.

twentieth century, the search for a technological fix was the earliest and most pervasive attempt to stem decline. Solving thorny problems of industry and depleted natural resources with high technology was especially seductive for politicians and affected workers because it seemed to offer a painless solution to what would otherwise be messy social and political conflicts. Yet the technological fix of taconite mining was never as simple or as easy as its proponents imagined.[2]

Inventing taconite iron ore was not a one-way street from idea to implementation. The original concept for processing low-grade taconite into valuable iron ore was well understood before Davis began working with taconite in 1913. Despite this body of knowledge, it took years in the laboratory to perfect taconite milling. And even that was not enough. Taconite also had to be economically viable in a competitive iron-ore market. As this chapter demonstrates, it took decades of work to perfect taconite and make it economically viable. To do this, Davis ultimately had to convince others of the threat from iron-ore depletion, change tax laws, and stave off foreign competition. In the end, the technological fix was anything but simple. But the rhetorical and conceptual genius of the technological fix was that it rarely acknowledged its full complexity.

Taconite was not alone in offering high technology and engineering as a solution for difficult problems of industrial decline and economic dislocation. In the United States, the middle decades of the twentieth century were saturated with an abundant faith in science and technology's power to overcome social and economic problems. Master scientists and engineers such as J. Robert Oppenheimer held an esteemed place in American life as wise visionaries who could help—or harm—the future. Yet the promises of technology were especially powerful in industrial regions such as the Iron Range that were already feeling the pinch of economic decline by the middle of the century. Technology seemed to promise the Iron Range and other industrial regions that they could maintain a prosperous industrial economy into perpetuity. The turn to taconite as a technological fix was arguably one of the most successful uses of engineering to reinvent a declining industry, yet it was not without problems. A close account of taconite's creation reveals the complex work required to create the technological fix and points toward its often unacknowledged and unintended consequences.[3]

Taconite before 1913

To understand the Iron Range's history in the twentieth century and why taconite was significant, it is necessary to differentiate between high-grade and low-grade iron ore. High-grade iron ore typically refers to ores that contain 50 percent or more iron content as they occur geologically. High-grade iron ores in the Lake Superior mining district generally take the form of hematite, a rusty red, gravel-like ore. The Mesabi Range contained scattered pockets of this high-grade hematite ore in places where non-iron minerals had been leached away over eons of geological time. The Lake Superior district's high-grade iron ores revolutionized the U.S. iron-ore mining industry at the turn of the twentieth century because they could be scooped from the ground and shipped directly to steel mills with little further processing. Thus, the steel industry often called them direct-shipping ores. With their high iron content, they could be added directly to the blast furnace.

Low-grade iron ore, on the other hand, generally refers to iron ore that contains less than 50 percent iron as it exists geologically. Several types of low-grade iron ore exist in the Lake Superior mining district, with the most common being taconite (in Minnesota) and jasper (in Michigan's Upper Peninsula). Both ores contain approximately 20 to 30 percent iron. Through industrial processing, low-grade ore can be concentrated to remove the non-iron material and create a manufactured product containing enough iron to be used in steel mills. Rudimentary forms of concentration involved sifting and washing iron ores to eliminate low-grade or non-iron material. But the more complicated industrial processing of the mid-twentieth century involved extensive crushing and grinding, followed by separation using magnets or flotation, with the iron-bearing particles finally concentrated and bound together into high-iron chunks or pellets.

For iron ores and many other minerals, technical developments in ore concentrating and milling had revolutionized the mining and minerals industry beginning in the mid-nineteenth century. In mining regions around the globe, inventive engineers and metallurgists developed techniques to efficiently and profitably recover metals from complex, low-grade ores. In gold-mining districts, for example, hydraulicking and the cyanide process allowed miners to recover ever-smaller traces of the precious metal from extremely low-grade ores. For zinc and copper, flotation allowed for profitable mining of low-grade ores in Australia and western North America. While the techniques varied depending on the ore and the location, the end result was a shift in the mining industry away from the older model of the lucky prospector and toward a rationalized, engineered process for turning low-grade ores into steady corporate profits. As mining historians Logan Hovis and Jeremy Mouat argue, the mining industry after 1900 was characterized by "the abandonment of low-volume, high-value, selective mining techniques and the adoption of higher-volume, nonselective methods that emphasized the quantity rather than the quality of the ore brought to the surface." Thus, the basic technical body of knowledge that Edward W. Davis drew from when he began working with taconite was well known by the early twentieth century.[4]

The principle of magnetic separation—using magnets to separate finely ground minerals of differing magnetism—was well established by the first decade of the twentieth century. A 1909 textbook on ore dressing described magnetic separation as a familiar and simple technology that was especially useful for separating strongly magnetic minerals such as magnetite. In practice, many iron-ore mines in New York and New Jersey used some type of magnetic concentration of magnetite ores by 1914. Scandinavian iron ores were also concentrated using magnets. Although the principles behind magnetic separation were well known, the use of magnetically separated iron ore was limited prior to the mid-twentieth century. As a textbook from the era noted, "as a matter of fact, concentrated ores are not an important factor to-day in the iron industry." Most iron ore used in the Great Lakes and southern U.S. markets was shipped directly, without concentration.[5]

Yet, not every attempt to use magnets to concentrate iron ores was commercially successful. Thomas Edison, for example, wasted much of the 1890s in a disastrous attempt to develop a magnetically separated iron-ore mine and mill in New Jersey. After he was forced from his leading role in electrical lighting, Edison devoted his energies toward a magnetic iron-ore separation venture in New Jersey, declaring that the iron mine would be "much bigger" than his previous work with electricity. Although Edison perfected an automated system for milling low-grade magnetic iron ore, his venture proved a costly failure. Buyers in the iron and steel industry did not want Edison's ore and his high costs could not compete with other ores, including the cheap Mesabi Range ores that began flooding the market in the 1890s. His New Jersey mill was eventually dismantled and he returned to his laboratory. His failed attempt to revolutionize the iron-ore mining industry through magnetic separation of low-grade iron ore remained a black hole in his otherwise fantastic career. Edison's biographers have offered a harsh judgment of the mining venture, calling it "an unmitigated disaster" that led him to lose "not only those years of his life but the entire fortune accumulated by his inventive labors," "a dusty, frustrating, dangerous, bottomless, and expensive pit," his "greatest commercial failure," and simply "Edison's folly." Despite the seeming simplicity of magnetic separation, there was no guarantee of commercial success in the field.[6]

Developments in copper mining, especially ore separation via flotation, were the most radical innovations that directly influenced later work with low-grade iron ore. Mining engineers and metallurgists working with copper ores developed techniques to mine and process vast quantities of low-grade ores in the late nineteenth and early twentieth centuries. The worldwide mining industry faced a crisis by the 1890s as the globe's rich and easily accessible ores petered out. The remaining ores were harder to reach and required more complicated milling than earlier mines. In an effort to overcome the limits of geology, metallurgists began experimenting with new milling techniques that could efficiently separate the valuable ore from the tailings. The most radical innovation that emerged from these efforts was the flotation method, in which a solution of slightly acidic and oily liquid was used to separate copper from tailings. Particles of copper adhered to the oil and floated to the top, where they could be skimmed off, while the tailings sank to the bottom for removal. Following successful implementation in the British Empire in the late 1890s, flotation techniques were adopted in western copper mines in the United States where they first were used to capture fine particles, or slimes, lost to older gravity separation methods. As mine and mill operators realized the efficiency of flotation, however, the technique soon replaced older separation technologies altogether in the western copper mines. Flotation not only revolutionized copper mining, but it also spurred mill operators and metallurgists to rethink the methods and business practices of mining, allowing them to exploit low-grade and complex ores on a scale never before seen in the mining industry. The technical knowledge gained from mining the enormous tonnages of low-grade rock, as well as crushing and grinding the ores, known as comminution, would prove essential for later developments in iron ore.[7]

Edward W. Davis's Early Work with Taconite

One man fascinated by the boom in new mining technologies was John Williams, a Mesabi Range landowner and regent at the University of Minnesota. Williams met Edward W. Davis, the new mathematics instructor at the School of Mines, during a routine tour of the school. During the visit, Williams told Davis about the mineral known as taconite, a flinty, gray rock common throughout northeastern Minnesota. Taconite had been passed over by miners in favor of the richer hematite. Williams was convinced, however, that taconite could be profitably exploited using new milling technologies. Not coincidentally, Williams's land on the eastern Mesabi Range contained large taconite deposits. He stood to gain an enormous financial windfall if an engineer could economically extract iron ore from taconite. Williams sent samples of taconite to the School of Mines and Davis soon began his experimental work. Davis's institutional home at the time—and throughout his long career—was in the Mines Experiment Station (MES). The University of Minnesota's School of Mines originally created an assay laboratory for testing gold in the late nineteenth century. Through contributions from local businesses, this small laboratory expanded into a larger Ore Testing Works and was renamed the Mines Experiment Station in 1911.[8]

Davis brought the taconite samples into his laboratory and recorded his initial observations about taconite. "It was a hard rock," he noted, "mottled in appearance, with a few narrow bands of darker material passing through it . . . After breaking the sample into small pieces, I found that most of them could be picked up readily with a hand magnet." Magnetism was an easily understood phenomenon and it was familiar to Davis from his previous work at Westinghouse and General Electric. Thus, it is not surprising that he reached for his hand magnet when testing the iron ore. As an engineer educated at the turn of the century, he would certainly have been familiar with earlier efforts to use magnets to separate magnetic iron ore, such as the use of magnetic separation in the eastern iron-ore regions and Thomas Edison's failed iron-ore venture in the 1890s.[9]

Although using magnets to separate finely ground iron ores was a well-established technology by the 1910s, Davis departed from existing iron-ore milling practices by adapting technologies developed in the western copper mines. Techniques used in the eastern magnetic iron-ore operations were not suitable for taconite. After testing iron ores from the East Coast, staff at the MES found taconite to be harder than eastern iron ores and that it contained less magnetite. As Davis noted, "magnetic ore shipped to [the Mines Experiment Station] from New York . . . was far simpler to concentrate than Minnesota taconite, since it contained over 40 per cent iron, was coarsely crystalline, comparatively soft, and easily crushed." Western copper mines, which were successfully mining and milling complex, low-grade ores by the 1910s, were a better model and many of Davis's early experiments drew on processes used in that industry, especially the process for comminution of large quantities of hard-rock ores. Davis soon recognized that taconite was more like the western hard-rock ores than eastern iron-ore deposits. It was very hard, meaning that it required the latest techniques in comminution, and its low percentage of magnetite meant that it would only be profitable if processed in enormous volume.[10]

Magnetic separation of taconite combined techniques developed for copper ores with existing magnetic separation technologies used in eastern iron-ore mines. For example, Davis created a magnetic log washer by adding an electromagnet to the existing log washer, which was a familiar machine for washing ores. He next arranged several machines into a series, known as a flow sheet, that moved taconite from rock to iron powder. Concentrated iron ore had to be agglomerated into harder masses that could be shipped easily and to prevent the ore from blowing out of the blast furnace. By the end of the 1910s, Davis had designed an effective system for crushing and grinding taconite and using powerful magnets to separate the magnetic iron ore.[11]

Following his success in the laboratory, Davis joined mining engineer and industrialist Daniel C. Jackling in a venture intended to transform his taconite experiments into a commercial mining operation. The venture was incorporated as the Mesabi Iron Company. Although he is little remembered today, Jackling was one of America's most famous industrialists in the early twentieth century. Rising from humble origins in Missouri, Jackling brought industrial processes of mass production to western copper mines and made an enormous fortune. By pairing with Jackling, the young Davis joined one of the world's wealthiest and most innovative engineers.[12]

Jackling and his financial supporters believed that lean iron ores such as taconite could be profitably mined using techniques similar to those that had proven successful in the western copper mines. As Jackling's financial backers wrote to potential investors, "there is nothing hitherto unknown, unusual or even untried in connection with the process employed." There was, in fact, a close collaboration between iron-ore mining on the Mesabi Range and low-grade copper mining in the western United States that has been overlooked by historians. Jackling is widely considered to be a pioneer of low-grade copper ore mining and milling ventures. Yet many of his insights into modern mining techniques were borrowed from open-pit iron-ore mines on the Mesabi Range. For instance, immediately before he commenced open-pit mining operations at

his Bingham Canyon mine in Utah, Jackling and his superintendent traveled to the Mesabi Range to observe how the new open-
pit mines used steam shovels. Jackling's early involvement in low-
grade iron-ore mining, along with the flow of technologies between
open-pit iron-ore and copper mines, suggests that Davis was but
one of many engineers developing new technological systems for
mass removal mining. As historian Timothy LeCain has argued,
the true significance of early-twentieth-century mining engineers
and metallurgists such as Jackling and Davis lies in their creation
of "a technological system for cheaply extracting huge amounts of
essential industrial minerals from the earth's crust."[13]

Hoping to join the bonanza of low-grade ore mining, Davis took
a leave of absence from the University of Minnesota in 1918 and
joined the investors to build a pilot plant in Duluth, Minnesota.
After three years of work in the pilot plant, the company built a
mine and processing plant at a remote location on the eastern
Mesabi Range in Babbitt, Minnesota. Although Jackling was named
president of the Mesabi Iron Company, his role was mainly limited
to management and financial oversight. The mine's day-to-day operations were handled by Walter Swart, a product of the Colorado
School of Mines and longtime colleague of Jackling. In essence, the
Mesabi Iron Company planned to import the crushing and grinding techniques used in western copper mining and then add magnetic separation to the process. But engineers quickly discovered
that taconite ore was much harder than western ores or eastern
magnetite iron ore. Although a successful flow chart for crushing,
grinding, separating, and sintering the ore was developed, it proved
more costly and energy intensive than the firm initially expected.[14]

The Mesabi Iron Company also encountered problems when it
tried to sell taconite on the iron-ore market. The firm expected that
its product would command a premium price because of its structure, which made it more efficient to use in blast furnaces, and its
low phosphorus content, which meant it could be used in Bessemer
converters. The Bessemer process did not remove phosphorus from
iron ore and thus required low phosphorus ore to produce high-
quality steel. In contrast, the open-hearth steelmaking process that
was being adopted in the United States allowed steelmakers to remove phosphorus and lessened the demand for low phosphorus
ores. Davis and his colleagues designed their mill to produce a final
product that assayed at 60 to 61 percent iron. Once the mill was in
operation, however, iron and steel companies were only interested

Employees of the Mesabi Iron Company ride a gasoline speeder at the first low-grade iron-ore mine in Babbitt, Minnesota, circa 1916. Edward W. Davis sits on the left. Although the Mesabi Iron Company failed in the 1920s, it was a forerunner of the taconite mines that would transform the Iron Range in the decades after World War II. Minnesota Historical Society Collection, HD3.22 p15.

in purchasing ore containing 64 to 65 percent iron. The entire process had to be redesigned to meet the higher standard. The underlying issue, which Davis and others did not anticipate, was that concentrated taconite contained more silica than other ores. Buyers in the steel industry realized that high silica levels meant they would have to waste more of the ore as useless slag and they were not willing to pay a premium for taconite. Without this premium price, taconite's complicated processing made it unprofitable.[15]

Thus, Davis and his colleagues at the Mesabi Iron Company found that their separation process was technically successful but was too expensive to compete on the iron-ore market. They were trying to manufacture iron ore for a steel industry that was undergoing rapid technological change. The Mesabi Iron Company's mill was built just as the decline of the Bessemer process reduced the

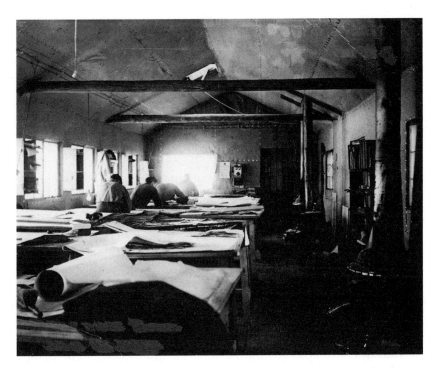

Engineers work in the drafting room of the Mesabi Iron Company's Babbitt location in 1921. Rough, camp-like conditions prevailed at the mine. Minnesota Historical Society Collection, HD3.22 r7.

demand for low phosphorus ores, one of taconite's chief advantages. The growth of open-hearth steelmaking and the rising cost of coking coal made low-silica ores especially desirable, a problem for the relatively high-silica taconite. By 1924 it was clear that the company could not profitably produce taconite and Jackling wrote shareholders that the company was closing its operations. The Mesabi Iron Company's mill closed in June 1924, after only two years in operation. Davis was hopeful that the mill would reopen shortly, but he would be bitterly disappointed. It took almost three decades before the Babbitt site reopened.[16]

Alarm over Natural Ore Depletion

After the failure of the Mesabi Iron Company, Davis returned to the Mines Experiment Station in Minneapolis and rededicated himself to taconite research. Because his attempts to turn taconite into a commercial venture had failed, Davis now turned to state politicians to fund his ongoing research. To gain support from politicians and the people they represented, he emphasized how taconite offered a technological fix to the problem of the imminent depletion of the Mesabi Range's hematite iron ore. The high-grade hematite iron ore originally mined on the Mesabi Range was being depleted at an alarming rate, he emphasized, and only heavy investment in low-grade taconite ore could prevent the collapse of the region's mining economy.[17]

When Davis first began working with low-grade iron ores in 1915, he faced an iron mining region that was anxious about the consequences of its tremendous success. The Mesabi Range had been a bonanza of easily accessible, high-grade iron ore for two decades. Yet the geological and technological factors that allowed the region to exploit so much high-grade iron ore so quickly also raised fears about depletion. If the ore could be removed on an industrial scale with such speed, would the gifts of nature hold out?

The Mesabi Range's high-grade ore was indeed being removed at a fantastic rate. Iron mining on the Mesabi Range removed so much ore so quickly because it was among the first instances of fully mechanized open-pit mining in the world. Unlike the slow, labor-intensive process required to dig ore from underground mines, gravel-like Mesabi Range hematite could be easily scooped from the ground using primitive steam shovels. Transportation of the ore was fully mechanized as well. Railroads ran directly into the pits. These railroads carried ore to docks on Lake Superior, where it was loaded onto boats for water transport to steel mills in Pittsburgh, Cleveland, and Chicago. The entire process was described by one historian as "an intricate ballet of large and complex machines." Mechanization quickly displaced human labor in the mines. During the decade from 1910 to 1920, the amount of ore produced annually per worker rose from 1,522 tons to 4,257 tons.[18]

At the time, engineers hailed the Mesabi Range mines as exemplars of conservation in the mining industry. During the late nineteenth century, mining engineers and operators in the United States were concerned that wasteful mining practices were costing millions of dollars each year in lost minerals. The Mesabi Range, in contrast, was held up as a model of modern conservation practices, which mining engineers at the time defined strictly as preventing costly waste. Full mechanization of the mining process in open-pit mines meant that the Mesabi Range mines wasted little of the ore, and certainly far less than the 10 percent of ore that was typically lost in underground mining. The combination of open pits and full mechanization meant that the ore body could be completely and efficiently removed.[19]

Mechanization shifted the mines' energy source from human and animal power to hydrocarbon-fueled shovels and locomotives. Historian Timothy LeCain notes that a similar process happened simultaneously in the open-pit copper mines of western North America. Here, too, it was mechanization, and especially the use of the enormous shovels developed on the Mesabi Range, that allowed for mineral exploitation in formerly unimaginable quantities. LeCain writes, "the steam shovel was nothing more than a device . . . to channel large amounts of concentrated hydrocarbon or hydropower energy into the previously slow and labor-intensive process of mining." Replacing humans with hydrocarbon-powered machines allowed miners on the Mesabi Range to extract ore at rates that were unimaginable decades earlier. From 1880 onward, iron-ore extraction in the United States grew quickly, rising 80 to 138 percent with each decade. But this huge extraction also raised fears of resource depletion.[20]

The enormous volume of iron ore being removed from the Lake Superior district was a source of both awe and dread. Observers were clearly impressed by the size of the giant open-pit mines, with many mines setting up observation stands so the curious could watch the growing man-made canyons. The giant open-pit mines offered a vision of the industrial sublime, defined by historian David Nye as "a man-made landscape with the dynamism of moving machinery and powerful forces" that "evoked fear tinged with wonder" yet "threatened the individual with its sheer scale, its noise, its complexity, and the superhuman power of the forces at work." Indeed, the observers' wonder was tinged with concern that removing so much ore meant that the district's mines would soon be tapped out.[21]

Alarm over resource depletion, especially in minerals, was hardly new in the early twentieth century. Economists, geologists, and miners were well aware of the problems of mineral depletion and had devoted extensive study to this issue and possible technical solutions since the middle of the nineteenth century. For instance, British economist William Stanley Jevons published *The Coal Question* in 1865. In this book, Jevons raised alarm over the possible depletion of British coal that was being mined in ever larger amounts for the nation's growing industries. He warned that depletion of easily accessible coal seams in Britain was unavoidable and would lead to national decline as coal prices rose. By the late nineteenth century, mining and engineering trade journals were filled with sophisticated analyses of mineral depletion and its consequences for various mining industries.[22]

One of the most significant indications of national concern over iron-ore depletion in the United States was a 1909 survey of the nation's mineral resources. The *Report of the National Conservation Commission* noted that iron ore and other minerals essential to American economic and military power would quickly run out unless the nation immediately undertook drastic conservation efforts. Regarding the nation's iron-ore reserves, the report described the high-grade ores of the Lake Superior region as quickly diminishing even though the district had been opened only seventeen years earlier. The report described a "remarkable rate of increase" in American iron-ore production during the decades from 1880 to 1909, with most of this increase coming from the Lake Superior district. Although there were an estimated 2.5 billion tons of high-grade iron ore remaining in the district, this was far below the 6 billion tons of ore that the United States was estimated to need before 1940. Worse, the district's production already was showing signs that the highest grades of ore were depleted. The iron content of ores mined on the Mesabi Range was falling steadily in the years before 1909. The report concluded with a pessimistic prediction for the region and the United States' iron-ore supply. "The present average rate of increase in production of high-grade ores can not continue even for the next thirty years, and that before 1940 the production must already have reached a maximum and begun to decline." A 1914 textbook on iron ores confirmed the report's findings, noting a "growing scarcity of high-grade ores in the Lake Superior district" and describing various efforts, such as washing and roasting, used to improve the quality of remaining ores. By the first decade of the twentieth century, then, the depletion of high-grade iron-ore reserves on the mighty Mesabi Range had reached a national audience.[23]

Like iron ore, copper was another essential mineral that faced a potential shortage in the early twentieth century. Skyrocketing demand for copper, essential to electrification, led many business leaders and politicians to worry about a copper famine at the turn of the twentieth century. Their concern was based on an older understanding of copper ore deposits as naturally occurring but rare. According to this theory, once all of the nation's valuable deposits had been discovered, supply would decline and prices would skyrocket. But this theory—which portrayed copper mining as "a giant national treasure hunt"—was overturned by the low-grade porphyry copper mines. These low-grade copper ores, ores that contained as little as 2 percent copper, were successfully being exploited beginning in the early twentieth century. In historian Timothy LeCain's words, the porphyry copper mines reimagined mines as "a product of engineering and technology rather than a natural treasure that had been discovered." The essential point is that Davis's alarm over iron-ore depletion was one voice in a much larger conversation over mineral depletion and its economic consequences. When Davis first sounded alarm bells over Mesabi Range depletion in the 1920s, miners and geologists had been considering the consequences of depletion for many decades.[24]

Davis publicized his argument that natural ore was running out from the 1920s through the 1950s, beginning with technical documents aimed at fellow engineers but soon expanding to a wider popular audience. Among his earliest arguments about depletion was a 1920 technical bulletin in which he described a coming crisis on the Iron Range as the natural ore supplies ran out:

It is, of course, recognized by everyone, that at some future date all the merchantable ore will have been removed from the [Lake Superior mining] district. This date is placed by various estimators at from 15 to 30 years hence . . . The history of the Lake Superior district will, undoubtedly, follow in a general way the history of most mining districts, and it is to be expected that some day, past or future, the district has reached, or will reach, the peak of its production. After the peak is passed, and it becomes generally recognized that the district is on a decline, the descent toward absolute depletion will be quite rapid.[25]

Davis argued that developing the Mesabi Range's low-grade taconite ores was the obvious solution to natural ore depletion. He warned that if development of taconite did not begin immediately, the entire Iron Range would be in jeopardy:

It will not only be necessary to utilize these [low-grade] ores in order to maintain the production of the district, but it will be necessary to begin the utilization of them in the very near future. If the furnace companies that have only a few years' supply of ore available are allowed to invest large amounts of capital in developing and bringing into production new mining districts, the Lake Superior region will immediately start on its decline. If, on the other hand, these furnace companies find that the low-grade ores can be utilized, the fact that the Lake Superior district is already in large production and is so well equipped to handle immense tonnages will cause them to contemplate seriously investing new capital in this district for the purpose of developing the low-grade iron ores. The development of such an industry on a large scale will extend the life of the district into the far distant future.[26]

Although these early warnings were aimed at a technical audience, Davis quickly realized that state politicians and residents of the Iron Range were the true targets for his concerns. They were the people who would be most affected by decline in the iron-ore mines and abandonment of the region. To reach this audience, Davis framed the problem of declining natural ore using homey metaphors that would be easily understood among audiences unfamiliar with geology. "The Mesabi iron range in structure is similar to a plum pudding," he told audiences:

In the past, the miners have been taking out the plums only. When the plums are gone the pudding is still left to be eaten. It may not be quite so palatable but by sweetening it up a bit, it can be made quite desirable. The "plums" are the iron mines from which ore is now being shipped containing less than 30% earthy [non-iron] material. The pudding is the iron formation which surrounds the plums and contains more than 30% of earthy material. There is probably several hundred times as much pudding as plums. These "plums" of iron ore will keep us going at the present rate for about 20 years. It will then be necessary either to mine the pudding or to quit mining iron ore entirely. On the other hand, if the pudding of low grade iron ore is used we shall still be "eating" for the next five or six hundred years.[27]

By presenting the iron-ore situation as a pudding metaphor, Davis spread alarm about ore depletion to a wide swath of the Minnesota public.

Unfortunately, archival records do not indicate the exact composition of the audiences for Davis's frequent speeches warning of depletion and promoting taconite. Yet it is clear that Davis used every opportunity to reach public audiences ranging from fellow metallurgists and mining engineers to state politicians to crowds of Iron Range residents. He used any available vehicle to promote taconite, at times writing technical bulletins, letters to the editor in newspapers, both on the Iron Range and in Minneapolis and St. Paul, speaking to social clubs, making formal presentations with lantern slides to state politicians, and even recording radio and television shows. He proved to be a masterful rhetorical promoter of taconite, subtly changing his tone to match the expectations of the audience. While Davis's technical bulletins were sober documents written for fellow engineers, when speaking to public audiences on the Iron Range he employed populist and masculine rhetoric to appeal to the miners. He urged one group in Hibbing to be "men enough" to meet the challenge of taconite, and played on Iron Range residents' antiurban sentiments by assuring them that they knew more about the region's problems than "big city economists, journalists and politicians." He also worked behind the scenes to shape how his allies discussed the depletion issue in public. Prior to an Iron Range politician's appearance on a radio show, Davis sent the state legislator a letter intended to guide the radio broadcast. He suggested that the legislator present an overall "impression of great things in store for the [Iron] Range," but also urged him to warn listeners of the possibility of the region's becoming a string of ghost towns by describing to listeners an imaginary "old timer telling how it used to be and what happened to it."[28]

Although Davis's arguments about natural ore depletion and the need for taconite as a technological fix fell on deaf ears initially, he gradually won converts over the decades. Concerns about natural ore depletion were muted during the 1930s when the Great Depression silenced the Iron Range's mines. But production skyrocketed during World War II and revived Davis's depletion argument. The fear of iron-ore depletion during wartime also roused alarm that iron ore was vital to national defense in an era of steel-clad armies.

In an alarming 1942 report to the United States War Production Board, Davis described a coming crisis in the iron-ore supply. The Mesabi Range could meet high demand in the near future, Davis wrote, but "new properties will be stripped each year" and "after

Experiments with taconite and a balling drum in 1943 at the University of Minnesota's Mines Experiment Station. Following the failure of the Mesabi Iron Company, Edward W. Davis and his staff spent the 1930s and 1940s trying to make taconite technically and economically viable. Minnesota Historical Society Collection, HD3.22 r4.

the larger properties are exhausted, it will become increasingly more difficult to produce the tonnage required from a larger number of smaller properties . . . It is shocking to realize that in a comparatively few years, the great steel industry dependent upon Lake shipments will find itself short of the necessary ore to meet emergency steel requirements." Davis insisted that his plan to switch from high-grade natural ore to low-grade taconite would "determine whether our future ore supply is to be secured from the Lake Superior taconites or from South America." The War Production Board rejected Davis's plan to begin taconite mining during the war for technical reasons: it thought that the taconite mines would not be operational before the end of the war. Yet the close connection between iron-ore depletion and national security would prove a boon for Davis's appeals on behalf of taconite.[29]

Wartime concern that American iron ore would run out continued into the postwar era. The years immediately after World War II were haunted by fears of resource exhaustion and worry that depleted minerals would lead inevitably to weakening American power in the Cold War. Just after World War II, President Harry Truman received a report estimating that the Mesabi Range's natural ore would be mined out in twenty-six years given present mining rates. Anxiety over resource exhaustion in this era was best illustrated by the so-called Paley Commission of 1952, which detailed critical shortages in many of the raw materials needed for the looming Cold War. By the late 1940s and early 1950s Davis's concerns about natural ore depletion had reached the nation's most influential policy makers and worries about resource depletion were high on the nation's agenda. Davis had long argued on behalf of taconite, but in the postwar years his claims were planted in the fertile soil of anxiety.[30]

Davis's argument that natural iron ore was running out was never uncontested. From its earliest articulations, Iron Range residents and steel industry executives challenged him as to whether the ore was really nearing exhaustion. Iron Range residents, many of whom earned a living in the natural ore mines, were some of Davis's most vocal challengers. Many believed that the depletion stories were a hoax and that the large mining companies knew of enormous deposits of natural ore but had not yet "discovered" them to keep them off state tax rolls. These arguments were not entirely baseless. Knowledge of existing ore bodies was far from complete and much of it was obtained from mining companies with powerful financial incentives not to "discover" existing ore until they were ready to remove it. Davis admitted that Minnesota's mineral tax situation largely prevented exploratory drilling to discover new ore on the Mesabi Range. "If [a mine operator] puts down drill holes and finds ore," he argued, "he is at once taxed and will continue to be taxed until he removes the ore . . . The tendency is, therefore, for the operator not to drill his land until he actually needs the ore. Then if he finds it, he immediately rushes in and mines it out just as quickly as possible." Natural ore was quickly being depleted, he worried, but he also could not say precisely how much ore existed in the ground.[31]

Other groups in Minnesota focused on this contradiction to echo the concerns of Iron Range residents. In 1947, a group called the Minnesota Emergency Conservation Committee disputed Davis's claims about the imminent decline of natural ore. The chairman focused on Davis's admission that, ultimately, it was impossible to know how much natural ore was left on the Mesabi Range: "How can you . . . state definitely one way or the other just what is or is not a true statement, in so far as our mine deposits are concerned, when by your own acknowledgement the State does not know how much iron ore we have in Minnesota? . . . Don't you think it is about time we cease making statements concerning iron ore deposits until we really know whereof we speak?" Speaking before Congress, U.S. Steel president Benjamin Fairless also disagreed with Davis's pessimistic reports about the rapid depletion of ore reserves. Fairless told Congress that "an adequate supply of iron ore in this country is primarily a production and cost problem. It is not a question of the exhaustion of the ore reserves of the United States." These criticisms challenged Davis's depletion argument, but they apparently did not particularly upset him. Ultimately, natural ore depletion was a means to an end for him. Raising concern over the possible exhaustion of natural ore was just one aspect of his larger project of promoting taconite. He was not attempting to decisively answer the depletion question, but rather to raise enough of an alarm that politicians and the public would be receptive to doing whatever was necessary to develop the taconite industry as a solution to the problem.[32]

Changing Taconite Taxation

In addition to emphasizing the threat from natural ore depletion and tying taconite to nationalism, Davis also attempted to change taconite's economic feasibility by lowering mineral taxes. Even if the Mesabi Range's natural ore was running out, taconite remained more expensive than the easily accessible natural ores. To change the cost equation and make taconite more competitive, Davis tackled Minnesota's iron-ore tax structure. If the state's tax laws could be changed to make taconite cheaper, he believed it would better compete with remaining natural ores.

The major tax problem for taconite was Minnesota's ad valorem iron-ore tax, which taxed the value of iron ore in the ground from discovery until its removal. The ad valorem tax created two major obstacles for taconite. First, it gave mine operators a financial incentive to remove the highest-grade ore as quickly as possible (avoiding the lean taconite) in order to reduce their tax burdens. Second, and more important, if taconite was eventually recognized as valuable iron ore it would then fall under the ad valorem tax. Essentially, the iron-ore tax laws were written for the rich natural ores, based on the assumption that a ton of rock in a mine represented a substantial (and valuable) amount of hematite ore. As a lean ore, only a small percentage of taconite was actually valuable magnetite and the bulk of the rock was worthless tailings. Taxing the full ore body of taconite in the mine would thus create a particularly high tax rate for taconite because much of what was being taxed was not iron but tailings. Davis's concerns about the ad valorem tax were piqued in 1940 when a Republic Steel executive warned him that steel companies would not invest in Minnesota because of high taxes. Davis claimed that the executive "told me to go back to my laboratory and have a good time with that taconite, but not to expect him to have any interest 'in that God-damned hard stuff or anything else out there in Minnesota until you get over the idea of taxing everything to death.'" Davis realized that the ad valorem tax seriously hindered taconite's development. As he explained, "at the university we realized that the application of the ad valorem tax to taconite would be disastrous . . . It seemed obvious to us that if taconite were to be utilized, some changes would be required in Minnesota's mineral tax laws."[33]

Beginning in the early 1940s, Davis launched a publicity campaign to change Minnesota's mineral tax code. He first turned to the steel companies as logical allies, but they proved unsupportive. The steel companies, faced with the practical problem of supplying ore to their blast furnaces, felt that foreign ore provided an easier solution than taconite and did not require the unsavory political work needed to change tax laws. Davis found more support among Iron Range politicians and residents. In July 1940, he launched a project to convince Iron Range residents to support a state law exempting taconite from the ad valorem mineral tax. The project was coordinated with several University of Minnesota faculty members and supported by influential newspaper editors on the Iron Range. Its fundamental goal was to convince Iron Range residents that eliminating the tax was in their best interest. The outline of the rewritten tax law took shape in the bar of the Fay Hotel in Virginia, Minnesota, on December 2, 1940. There Davis met with a group of Iron Range politicians who said they would support a new law written by "the university people."[34]

Although several Iron Range politicians supported the proposed change to the mineral tax law, many Iron Range residents were opposed to preemptively decreasing taxes on a future taconite industry. To resolve these concerns, Davis spent much of 1940 and

1941 speaking on the Iron Range and encouraging residents to support taconite's exemption from the ad valorem tax. Describing his approach in these speeches, he said: "I told the audience that the future of the Mesabi Range 'lies in your ability to use low-grade materials,' but that this would never come about without a change in the ad valorem tax law." In 1941, Iron Range politicians introduced the revised tax law in the Minnesota House and Senate. To ensure that the bill moved smoothly through legislative committees, Davis put his laboratory staff to work on behalf of influential committee members, preparing statistical summaries and formulas for legislators. The law passed easily and taconite was excluded from the ad valorem tax beginning on April 23, 1941.[35]

Yet, rewriting state tax laws opened up another Pandora's box of problems. Because he argued that taconite production was essential for the state of Minnesota—not just the Iron Range—Davis needed to convince people across the state that they too had a stake in taconite's success. To accomplish this, he spent part of his time in the 1940s and 1950s promoting taconite among Minnesotans beyond the Iron Range. Among the most successful publicity efforts were his frequent visits to Minnesota classrooms. Historian James Ross remembered one such visit, recalling, "I was in fourth grade when [Davis] stood before me and my friends. He gave thousands of the state's school children little kits of taconite pellets, vials of lean ore and magnets, and taught them how to spell taconite and to duplicate his early experiments with magnetic separation." The Mines Experiment Station also offered a "taconite kit" via mail. The kit offered four samples of taconite in stages from rock to crushed powder to manufactured pellet. Along with the samples came instructions for re-creating experiments at home by, for example, crushing the rock with a hammer and picking up the iron with a magnet or mixing the iron powder with flour and baking it in the oven to create a taconite pellet. The kits were advertised during a television special that encouraged viewers to "do this in your own home" and "see for yourself how simple this process is that means so much to Minnesota's iron mining future." Taconite, Davis argued, required financial help from the state in the form of lower taxes, but achieving lower taxes demanded a statewide effort to justify taconite to residents with little connection to the Iron Range or mining. By connecting with people around the state, Davis guaranteed statewide support for the revised tax laws.[36]

Like the natural ore depletion argument, Davis's efforts to change the tax code were sharply criticized by powerful figures in Minnesota. In a personal letter, Minnesota Governor Orville Freeman criticized Davis for his frequent editorials calling for changes in taxation. "The premise of your article seems to be that we will need to satisfy from here to eternity with some kind of guaranteed fool-proof system the fears of people who might be investors in our state," Freeman wrote to Davis. "If they are concerned about some distant development which might affect them tax-wise, the truth of the matter is that there can be no such assurances forever." Yet the opposition of powerful politicians did little to slow taconite's development. This was, in part, the result of Davis's political acumen, for his promotion of taconite gained enough public support that it could overcome such criticism.[37]

Despite his role in lowering tax rates for mining companies, there is little evidence that Davis was personally opposed to corporate taxes. Indeed, he was aware that business executives could use the ad valorem exemption to argue for lower corporate taxes across the board. In a personal letter, he wrote that he was cynical of one steel executive's reasons for supporting the tax reduction: "It is sort of a hobby of [the executive] to blame everything on this tax policy although his company pays mighty little tax in Minnesota."[38] Davis's general belief was not that corporate taxes should be low, but that they should be predictable. Steady, predictable taxes were important, he argued, "so the operator can figure his costs accurately and give positive assurance that he can pay back the money he has borrowed to build his taconite plant." Just like the machines in a laboratory, state tax laws needed to operate in a rational, predictable manner to allow for proper calculation and engineering.[39]

Davis also promoted tax cuts as a solution for Iron Range residents worried about jobs and the local tax base. He described how a tax code conducive to corporate investment in taconite would help Iron Range miners. He focused on the issue of employment, suggesting that increased business investment in taconite could guarantee long-term, high-paying jobs in the mining industry. When speaking to Iron Range communities, he emphasized that their future hinged on taconite. "The long-range prosperity of the people of the Mesabi Range is dependent almost entirely upon the utilization of taconite," Davis told crowds. "The one important problem before the present generation is to get a taconite industry established." Davis frequently argued that taconite production required two or three times the manpower per ton of ore produced than natural ore. If a taconite industry began operation, he told audiences that the Iron Range should make plans for large population increases to meet the high labor demand. He also tried to alleviate the concerns of municipal officials who worried that lowering taconite taxes would decrease city revenue. Suggesting the motto "Forget the tax dollar, give us the payroll dollar," he asked Iron Range communities to compromise with the mining companies to guarantee continued employment through low taxation.[40]

Davis's predictions for an employment boom from taconite were not uncontested. Executives at other Minnesota mining companies, who perhaps knew best the employment requirements of mining operations, were skeptical about his employment claims. In a personal letter, a U.S. Steel executive warned that the statistics about taconite's increased labor needs were totally unproven: "While the industry had long been talking about the necessity of mining three tons of crude for every ton of taconite concentrate produced, we now know that this did not mean three times as many miners for an equal tonnage. Actually, no one had really figured out how many people there would be." While taconite production required more rock to be mined than natural ore, more of the work required in taconite mining and processing was automated, making it difficult to determine future employment needs. Another mining executive warned that publicity about taconite's positive economic benefits was creating unrealistic expectations in the Iron Range communities. He argued that touting taconite's economic benefits would likely "impart a kind of boom psychology" in the area and "feed the flames of enthusiasm on the part of hospital commissions and school boards for substantial expansion of their facilities." Once again, these criticisms did not slow the enthusiasm surrounding taconite.[41]

Conclusion

Finally, after emphasizing the depletion of natural iron ore and changing the tax code, Davis's efforts were rewarded when two steel companies agreed to build a taconite plant. In the late 1940s and early 1950s, Armco and Republic Steel invested in taconite through their jointly owned company, Reserve Mining. Other steel firms quickly moved into the taconite business without Davis's help. Yet all the taconite producers benefited from the technical research and the political and economic support for taconite that Davis had helped to create. By the early 1950s, Oliver Mining built a pilot

Silver Bay, Minnesota, shown here in the winter of 1951–52, was an isolated outpost on the North Shore of Lake Superior prior to the construction of the E. W. Davis Works. Minnesota Historical Society Collection, HD3.27 p37.

plant to experiment with taconite production at Mountain Iron, Minnesota, and Erie Mining (subsidiary of Picklands, Mather and Company) constructed another experimental plant near Aurora, Minnesota. The taconite boom had begun on the Iron Range.[42]

Davis's personal connection continued with Reserve Mining Company. In 1951 he took a leave of absence from the University of Minnesota to assist with the construction of Reserve Mining's plant on the North Shore of Lake Superior. The plant processed its first taconite pellets in 1955, and Davis retired that year to work as a consultant for the company. Reserve Mining was so grateful for his hard work and dedication that it named the plant in his honor: the E. W. Davis Works. Davis and his wife soon moved from Minneapolis to a new home on a hill overlooking the plant named for him.[43]

The plant was a massive affair. Taconite was mined at the Peter Mitchell mine near Babbitt, Minnesota. There, workers blasted out

A TECHNOLOGICAL FIX?

By the 1960s, Edward W. Davis had retired from the University of Minnesota and moved to a home in Silver Bay overlooking the Reserve Mining Company's taconite mill, named the E. W. Davis Works in his honor. Minnesota Historical Society Collection, por 5957 p5.

piano-sized chunks of the hard-rock ore and loaded them onto rail cars for the trip south to the shore of Lake Superior. At the other end of the rail line was the E. W. Davis Works, where the taconite boulders were crushed and ground to the fineness of flour, the iron was separated from the siliceous rock, and the iron ore was baked into pellets. Staffing the Reserve Mining operation was an equally massive affair. The taconite operations used previously overlooked ore bodies, which meant that they were often located far from the existing Iron Range towns. The Reserve Mining Company ultimately built two entirely new towns for their workers, many of whom had to be brought in from Kentucky and New York because of the specialized skills required in hard-rock iron-ore mining. To staff the lakeside plant in Silver Bay, Minnesota, Reserve Mining hired an army of several thousand construction workers who lived in a nearby barracks and ate in a mess hall.[44]

Reserve Mining Company's E. W. Davis Works was hailed in the early 1960s as a remarkable feat of engineering. This photograph shows a tunnel under Highway 61 that moved taconite from one stage of processing to the next within the massive mill. Photograph by Durant Barclay Jr., circa 1960. Minnesota Historical Society Collection, HD3.27 p39.

Edward W. Davis and Reserve Mining Company executives shake hands in front of the first train car that hauled taconite from the mine in Babbitt to the mill in Silver Bay, circa 1963. Minnesota Historical Society Collection, por 5957 p6.

Reserve Mining Company's E. W. Davis Works was a success in the years immediately following its opening. Initially designed to produce 3.75 million tons of taconite per year, the plant was producing 6 million tons annually by 1961 as a result of greater than expected efficiency and several expansions. By 1970, pelletized iron ores such as taconite made up three-quarters of all iron ore used in the U.S. steel industry. Throughout the 1970s, the plant supplied 15 percent of the U.S. iron-ore supply.[45]

The taconite industry thus became the first great fix for the Iron Range's economy in the twentieth century. Instead of gradually declining as the high-grade natural ore ran out, taconite used high technology to seemingly obliterate natural limits. Thanks to the miracles of technology, it appeared that the low-grade ore would last for centuries. The problems of decline and depletion had seemingly been overcome thanks to modern science and engineering: a perfect example of the technological fix at work.

The Reserve Mining Company was one of several firms that invested in taconite mining in the 1950s and 1960s. This photograph shows workers building Erie Mining Company's taconite mill near Hoyt Lakes, Minnesota, in 1955. Photograph by Gordon Ray. Minnesota Historical Society Collection, I.198.73.

The public's view of taconite and Davis was equally positive. At the end of his long career, Davis was heralded as the engineer who had saved the Iron Range. One typically glowing newspaper story described him as "partly Thomas Edison and Carl Sandburg." Another reporter claimed that the large taconite plants "stand as colossal reminders that there is little in the world as good as a good and dedicated scientist like E. W. Davis." Academic institutions also recognized Davis's accomplishments; he received honorary degrees from Purdue University and the University of Minnesota, as well as several honorary awards from the iron and steel industry. In sum, he stood as a personification of the era's great faith that science and engineering could overcome the intractable problems of industrial decline and natural limits.[46]

Reality on the ground was more complicated. Taconite ore did resolve the problems of depletion, but it was hardly a perfect cure

A TECHNOLOGICAL FIX?

for the Iron Range's economy. For example, the high pellet production achieved at the Reserve Mining plant created problems because it was producing more iron ore than blast furnaces could use. The ore was stockpiled and, as the stockpile grew, Reserve Mining was forced to cut back operations in 1958. Eventually, the plant was shut down every fourth week to prevent oversupply.[47] And the success of the taconite plants came at the expense of the remaining natural ore mines, which faced declining demand for their products and began shutting down. The workers and communities surrounding such mines suffered significant economic hardship because they could not compete with taconite. Entire mining districts, especially in Michigan and Wisconsin, collapsed. Davis was far less sympathetic to the economic dislocations that taconite caused in other mining regions. Reflecting on the success of taconite, he wrote, "the great improvements in smelting practices, brought about by the availability of pellets made to specification, has caused a corresponding revolution in the ore fields. Mines that have been producing high-grade ore for more than half a century are being abandoned, not because they are depleted, but for lack of a market."[48] While taconite's benefits were obvious (they could be seen by anyone who drove past the enormous pelletizing plants and their surrounding towns), the costs were much harder to see. They were dispersed throughout the region and often felt far from the plants. Sustaining a taconite industry required changes to the tax code and the Iron Range, and in later years the environmental costs of taconite would be dramatically emphasized and publicized.

Creating a Tax-Cut Consensus on the Iron Range

The 1964 Taconite Amendment

Creating the technical process for mining and milling taconite was a major accomplishment. Engineers such as Edward W. Davis worked for decades to bring a taconite industry to life in northeastern Minnesota. Yet technology alone could not guarantee that taconite would revitalize the Iron Range's depressed postwar economy. Taconite producers claimed they needed more financial assistance to be competitive in an increasingly global postwar economy. Davis and his colleagues had revised Minnesota's mineral tax laws in 1941 to make taconite more economically competitive. Yet once taconite plants were up and running, it was revealed that they might need even more financial assistance to remain viable. To do so, taconite's supporters and their political allies passed an amendment to Minnesota's constitution that limited taxation on taconite producers.

Debate over the so-called taconite amendment began as a technical matter involving mining engineers, economists, and tax-policy experts. But during the early 1960s the dispute intensified as liberal and conservative politicians used taconite taxation as a divisive wedge issue that forced Iron Range residents to wrestle with competing political identities: the liberal working class versus the conservative taxpayer. Throughout the first half of the 1960s, the taconite taxation debate was among the foremost political issues in Minnesota. It eventually split the state's Democratic-Farmer-Labor (DFL) Party and offered conservatives entrance into the historically liberal Iron Range. The debate culminated with passage of an amendment to Minnesota's constitution in 1964 that ensured continued low taxes for taconite and limited any future tax increases, a dramatic reversal of earlier attitudes on the Iron Range that emphasized high taxes on mining companies as recompense for the removal of the state's natural resources.

Tax policy is an important part of understanding the postwar history of the Iron Range. But taxes and tax policy also offer valuable insights into the wider course of twentieth-century U.S. history. Taxes, especially corporate taxes, vividly illustrate a society's common ideas about the appropriate relationship between corporations and the state. Through tax policy, a society puts down in black and white its understanding of what, if anything, companies owe to the larger political body. These policies obviously cannot reflect the multiplicity of ideas within a population, but they nonetheless offer historians one way to reconstruct past attitudes about the relationship between business and government.[1]

If tax codes reflect social beliefs, then those moments when tax policy is substantively changed deserve special attention. In the twentieth century, politicians have endlessly tweaked the tax code for political advantage, thus creating the Byzantine modern U.S. tax system, which is filled with loopholes, credits, and deductions meant to appeal to specific groups that were politically important at a particular moment in the past. The tax code is like an archaeology of past political concerns, where the needs and wants of past interest groups are semipermanently sedimented into the present. Nevertheless, if the nuts and bolts of the tax code are driven by political imperatives, the larger terrain of possibility for taxation— what is and is not thinkable to politicians at any given moment—is a reflection of widespread ideas concerning the appropriate scale and role for the government in the lives of citizens and the economy.

Historians have powerfully demonstrated how the tax code reflected deep-seated beliefs about business, labor, race, and gender throughout American history. In the early nineteenth century, for instance, Robin Einhorn has shown how the U.S. tax code reveals slavery's central role in shaping American government. Alice Kessler-Harris has similarly demonstrated how the income tax reveals the gendered assumptions about economic citizenship and equality that ran throughout twentieth-century U.S. history. In the second half of the twentieth century, Julian Zelizer has noted how the federal tax code became the primary vehicle for government intervention in the economy. Clearly, tax historian W. Elliot Brownlee is correct to suggest that "tax politics was always an important vehicle for the expression of both national values and . . . underlying social and ideological conflicts."[2]

The history of the taconite amendment illustrates how shifting attitudes toward corporate taxation reflected deep-seated changes in the relationship between corporations and citizens in the decades after World War II. The Iron Range was not alone in reaching a consensus to cut corporate taxes in the postwar decades. As Zelizer notes, corporate tax policy and individual tax policy were linked in the postwar decades by "the guiding assumption . . . that increased revenue would be produced automatically from economic growth." Congress thus put downward pressure on federal corporate taxes throughout the 1960s, giving corporations a large investment credit in 1962 that caused the highest tax rate for corporations to decline "from 70 percent in 1964 to 36 percent in 1986." Political scientist Cathie J. Martin describes the overall pattern of declining corporate taxation as a percentage of all taxes paid in the twentieth-century United States as a long-term "shifting of the burden" of taxation away from corporations and onto individuals in the form of payroll taxes. In the early 1950s, corporate taxes provided approximately 30 percent of all tax revenue at the federal level. By 1983, this amount declined to 8 percent. Over the course of the postwar decades, corporations paid for a smaller and smaller share of government, with individuals taking up the slack via payroll deductions. Decisions to lower taxes for taconite producers on the Iron Range were one important piece of a larger national story about the shift from corporate taxes to individual taxes in the late twentieth century.[3]

One key significance of the taconite amendment is that it highlights one of the great wedge issues of post–World War II U.S. history:

the tax revolt and its role in the long, slow decomposition of the New Deal order and creation of a New Right majority by the last decades of the century. Although the reasons for this political shift were multifaceted, one of the main levers in prying apart the New Deal coalition was taxation. Beginning as early as the 1950s and 1960s, conservatives sought to undermine labor's connection to the Democratic Party by invoking another identity, the taxpayer, among blue-collar industrial workers. American factory workers were not just rank-and-file union members who should blindly follow the political decisions of union leaders, this rhetoric implied, but also middle-class taxpayers who deserved value for their money from the government. The idea of citizens making demands on the government as taxpayers was not new in the 1960s. Historians have found examples of taxpayer movements stretching back into the nineteenth century. Nonetheless, the contours of the blue-collar taxpayer revolt were unique to the postwar era, made possible by broad postwar affluence and a Cold War political climate that stifled expressions of class-based identity. More important, when the financial stability of blue-collar workers began eroding owing to deindustrialization and increasing global competition in the middle decades of the twentieth century, the political rhetoric of the taxpayer took on increasing importance in many industrial communities.[4]

Like workers in many other industrial regions of the United States, Minnesota's iron-ore miners benefited from unionization and collective bargaining rights during the postwar decades. Yet the economic security of union mining jobs came only after several decades of intense struggle to unionize iron-ore miners in northern Minnesota. Early miners on the Vermilion and Mesabi ranges, many of whom were European immigrants, went on strike in 1907 and again in 1916 to protest corrupt bosses and poor working conditions. These strikes were brutally repressed by the mining companies and their failure stifled union organizing among miners for several decades. During the 1910s and 1920s, the mining companies used a combination of tight labor control and limited welfare programs for workers to prevent further union organizing activities. With the national upsurge in union organizing during the 1930s, however, there was a renewed push to unionize Minnesota's iron-ore miners. The Steel Workers Organizing Committee (SWOC) began organizing iron-ore miners in the 1930s as part of its larger effort to build a comprehensive union in the steel industry. Yet

SWOC's attempts to organize Iron Range miners were initially unsuccessful. Minnesota's iron-ore miners were geographically distant from the steelmaking centers in the lower Great Lakes and had very different job classifications from workers in the steel mills. In the words of historian Donald Sofchalk, "many iron miners felt that the SWOC hierarchy treated [the Iron Range] locals as stepchildren, slighted their interests, and considered them to be of little more than peripheral importance in relation to the steel union as a whole." The mining companies also emphasized iron-ore mining's distinction from steelmaking in an effort to prevent industry-wide unionization. Concerns such as these hampered union organizing on the Iron Range throughout the late 1930s.[5]

The tight labor market of World War II and federal officials sympathetic to organized labor finally tipped the Iron Range miners toward unionization in the 1940s. Demand for iron ore and mining laborers increased sharply in 1941 and 1942. No longer worried about retaliation from mining companies, miners were eager to join the United Steelworkers of America (USWA; the SWOC was integrated into the newly formed United Steelworkers of America in 1942). By 1943, the large open-pit mines of the Iron Range were under the organization of the USWA. Iron Range miners, like so many other industrial workers in the United States, entered the postwar years buoyed by prosperity and a new sense of union-backed economic security. Yet these successes led to new political considerations, including taxation.[6]

It was only in the decades after World War II that mass taxation became a political issue in American life, largely because this was the first time that income taxation extended to the majority of the American people. Although the Sixteenth Amendment to the U.S. Constitution permitted a national income tax beginning in 1913, taxation in the early decades of the twentieth century rarely reached industrial workers. It was only with the onset of World War II, which spread prosperity to many industrial workers and put enormous fiscal strains on the federal government, that taxation affected most industrial workers. Wartime taxes were portrayed as a patriotic sacrifice, however, so it was not until the decades after World War II that many industrial workers and poor Americans began thinking about the long-term implications of mass taxation. As historian W. Elliot Brownlee puts it, during the postwar era "mass taxation had replaced class taxation," with 90 percent of all wage earners now submitting income-tax returns.[7] Politicians also learned important lessons about the impact of sustained mass taxation in the post-war decades. For liberals, the enormous sums of tax money flowing into government coffers meant that government could manipulate the national macroeconomy through taxation. Conservatives, too, took away important political lessons from mass taxation, realizing that year after year of paying income taxes changed many citizens' attitude toward government. Sustained mass taxation meant that so-called big government and high social spending were not just ideological questions, but pocketbook issues for most Americans.[8]

On the postwar Iron Range, the new politics of taxation ran headlong into an established tradition of activist local governments funded by high taxes on mining companies. By the postwar era, the Iron Range was among the most staunchly liberal regions in the United States. Two generations of active unionism and hostile relations between mining corporations and miners culminated in wholehearted support for Minnesota's DFL Party, with many Iron Range politicians supporting the party's farmer–labor wing. Sharp antagonism to the mining corporations—portrayed as grasping, greedy corporations based outside the region—permeated the DFL's Iron Range rhetoric. A *New York Times* reporter visiting the Iron Range in 1946 noted the region's commitment to liberalism: "the Iron Range is still the stronghold of the nationalistic Democratic-Farmer Labor party." Given this political dynamic, Minnesota Republicans largely wrote off the Iron Range as a lost cause in the middle of the century.[9]

Taxation changed the region's political calculations by the 1950s and 1960s. The appearance of the taconite amendment in the late 1950s offered a unique wedge for Republicans to stake a political claim in northeastern Minnesota. Republicans portrayed the amendment as a tool for growth in an industry that was reeling from global competition. Perhaps nowhere was the issue clearer than on the billboards that mining companies erected along Iron Range highways during the 1961 amendment controversy. The billboards asked Iron Range drivers, "Mining Jobs or Taxes, Which Help You the Most?" Narratives of culture war have gained widespread salience since the 1960s, but the battles over the taconite amendment illustrate simultaneous fiscal wars that worked alongside the politics of cultural outrage. More than the culture wars, though, partisan fiscal debates directly affected how government could operate. By cutting government revenue or shifting it away from corporations and toward individuals, the fiscal revolt transformed government, on many levels, into a leaner entity that emphasized efficiency and was more directly responsible to taxpayers than the liberalism of the World War II era.[10]

A second major significance of the taconite amendment is that it was an early example of the tax-incentive growth policies that spread throughout the United States by the end of the twentieth century. The underlying rationale of tax-incentive growth policy was to use targeted tax cuts to stimulate specific sectors of the economy. Examples such as the taconite amendment demonstrate how these policies came into widespread use in the postwar era, beginning with the focused local- and state-level cuts seen in the taconite industry. By the 1960s and 1970s, the American tax code, from the federal to the local level, was riddled with these loopholes aimed at stimulating various politically influential sectors of the economy. While the economic effects of these tax cuts are debatable, the outcome for liberalism was undeniable. The belief that cutting taxes in a given industry could stimulate the economy was deeply corrosive to the liberal state. The antipathy toward taxation that holds sway in American public life at the beginning of the twenty-first century has many precedents, but industrial tax incentives such as the taconite amendment are an important part of its family tree.

Taxation on the Iron Range before the 1960s

Consensus that taconite firms on the Iron Range needed a tax cut in the 1960s was a striking reversal of the region's past attitudes toward corporate taxation, especially the taxes that Iron Range cities imposed on the mining firms. Throughout the early twentieth century, the Iron Range stood out as a haven for working-class radicalism in the United States. Practically, this radical sentiment was channeled into high corporate taxes on mining companies. Fueled by a conviction that mining firms were greedy, profitable corporations, Iron Range residents believed that their towns had to take what was rightfully theirs through taxes. Thus, support for a tax cut for taconite firms in the 1960s illustrates how economic decline undercut older patterns of anticorporate radicalism in the postwar decades.

Taxation of Minnesota's mineral land and mining companies had a complicated early history. When iron-ore mining companies began operating in northeastern Minnesota in the 1880s, mine owners moved quickly to limit their tax obligations. In 1881, one mine owner spearheaded passage of a state law that removed all property taxes on mining corporations and replaced them with a production tax of one cent per ton of iron ore shipped. The tax was meant to ensure that mines could not be taxed on their large capital investments before they began producing ore. Mines were free from property taxes until 1897, when the earlier law was declared unconstitutional. From 1897 to 1914, mines were taxed on the same basis as all other property in Minnesota. After acknowledging that proper valuation of mineral property was especially difficult, the state enacted special—and relatively high—property taxes on mines in 1913. By the 1910s, many Iron Range villages also established special taxes on mines that greatly increased the revenue of the towns and taxed mines at a high rate compared to other businesses. A pure resource-extraction industry, mining posed thorny problems for taxation and was a flash point for competing ideas about the relative responsibilities that corporations, communities, and the state had to one another.[11]

The idea that mines should pay higher tax rates than other businesses resulted from the natural heritage theory of taxation, prominent among politicians in the first decades of the twentieth century. According to this theory, Minnesota's iron ore was a unique "natural heritage" of wealth for the state and any company removing that wealth should pay high taxes in recompense. Describing the tax policies in 1932, one economist wrote:

> the special taxes levied on the mining industry were the outgrowth of a general opinion in certain parts of the state that the mining companies were not required to pay as much as they should under the general property tax then in force. The natural heritage and diminishing value theories, and especially the belief that foreign corporations made large profits from Minnesota ores, probably had much to do with creating a public opinion in favor of special mine taxation.

Iron Range residents were not alone in their belief that taxes should be used to penalize corporations for taking wealth that rightfully belonged to the people. At the federal level, too, tax policy during the Progressive Era emphasized high tax rates on corporations and the wealthy as punishment for those making what were deemed excess profits.[12]

On the Iron Range, the natural heritage theory of taxation and anticorporate radicalism were personified in populist mayor Victor Power of Hibbing. Power was a miner and attorney in Hibbing during the early twentieth century. In 1913, he was elected mayor and proceeded to raise local property taxes on the mining companies to dramatically higher levels. He raised the village taxes on mining companies approximately 1,000 percent during the late 1910s. He used the high taxes to hire seasonally unemployed miners and build impressive public buildings in the town. Under fire from the mining companies and state politicians in St. Paul, Power defended his high-tax, high-spending regime in deeply populist language, telling voters, "Did we do wrong . . . to take money from [Charles] Schwab to prevent him from breaking the bank of Monte Carlo?" As Power's language made clear, Iron Range residents saw the high taxes as rightful recompense for the loss of valuable natural resources to rich and powerful business owners who already made what Iron Rangers thought were excessive profits.[13]

The discovery of iron ore underneath a portion of Hibbing and the subsequent relocation of many buildings to make way for expanded open-pit mining also enabled the city to justify expensive new public buildings. A body of iron ore lay underneath a section of Hibbing known as the "north forty." By the 1910s, mining operations had moved so close to this section's existing buildings that foundations were shifting and the neighborhood was surrounded on three sides by growing open-pit mines. In response, many buildings in Hibbing's north forty were razed and others were relocated to a new site. Relocating buildings to the new location provided a convenient justification for expensive new buildings—paid for by taxes on the mining companies that had required the move to expand their mines—in the new central section of Hibbing.[14]

By the 1930s, taxation of mining firms was hotly debated on the Iron Range. Residents and civic leaders pushed for higher taxes and mining companies sought tax relief. From the perspective of Iron Range residents and their elected representatives, taxation was the only weapon to challenge the power of the mining companies. As a sociologist working on the Iron Range at the time wrote, "through taxation [Iron Range city leaders] could acquire not only poor relief for labor but subsidies for struggling business enterprises; not only necessary conveniences for the public good, but luxuries." The Iron Range's high municipal tax system reached a crisis during the Great Depression when the economic collapse drastically

Hibbing High School around 1940. Hibbing's opulent high school symbolized the populist policy of taxing mining firms heavily to pay for luxurious municipal amenities in Iron Range cities. Minnesota Historical Society Collection, MS2.9 HB5.2 r23.

cut mining operations. The crisis occurred because the mines were shipping little to no iron ore, yet the taxes levied by local cities and school districts on the total amount of ore left in the ground continued unabated. Mining companies were upset at this continued high taxation despite the terrible economic conditions for their businesses. Even sympathetic observers saw the Iron Range's lavish municipal spending, paid for by high taxes on mining firms, as problematic amid widespread rural destitution. In a personal letter to Eleanor Roosevelt, for instance, government reporter Lorena Hickok described Depression-era Hibbing as "a sort of story-book town" where the expensive high school contained a clinic "that would do credit to the most perfectly equipped metropolitan hospital." Citing the town's public buildings, Hickok concluded that there was "darned little destitution up here" when compared with other rural areas she had visited.[15]

Thus, the Iron Range's system of high mining taxes was under assault by the World War II era. Criticism of high municipal taxes on the Iron Range reached a national audience in 1944 when a muckraking article in *American Mercury* mocked the city of Hibbing on the Iron Range. Hibbing, readers were told, was a "razzle-dazzle village" that gouged the mining companies through taxes and wasted the money on lavish public expenditures with no view toward the city's long-term fiscal health. The article expressed mocking wonder at Hibbing's public facilities, fawning over the city's $4-million high school, quarter-million-dollar city hall, and lavish public recreation building: "[Hibbing] is everything a mining town is not supposed to be, a hundred-million-dollar spot set in the backwoods." The article did not, however, suggest that these public facilities were beneficial in a dreary northern mining town or recompense for the removal of so much natural wealth. Instead, they were portrayed as essentially wasteful public expenses, turning the city into nothing more than "a wonderland of municipal doo-dads." Reporter Nathan Cohen blamed Hibbing's wasteful approach on Victor Power. Cohen described the popular nine-term mayor as "the whoop-de-doo boy of the range country, the original boondoggler, a political virtuoso who was a Huey Long to the mine owners, an Abraham Lincoln to the miners and Santa Claus to the building contractors and equipment salesmen." In Cohen's view, however, the city's residents ultimately were responsible for continuing the tradition of high taxation after Power left office. They created a community where the mining corporations were subject to high taxation and the money was frittered away on municipal opulence, a result, Cohen believed, of "a paw deep in the tax pocket of the mine operators who pay the bill . . . They tax to the limit, and spend to the hilt." To *American Mercury*'s national audience, Hibbing—and by association the other cities of the Iron Range—epitomized the problems and waste associated with high taxes on productive industry.

Despite Cohen's attempt to paint Hibbing as a corrupt mining town drunk on tax revenue, the voices of local residents hinted at an alternative understanding of the Iron Range's mining tax system, which used high taxes as a form of public welfare in a one-industry town dependent on the notoriously fickle iron-ore market. It was true that the city collected large tax revenues from the mining industry. From 1914 to 1944, mining companies in and around Hibbing paid $64 million in municipal taxes. But one resident explained that the high taxes were a logical response to the extractive nature of iron-ore mining and meant to take advantage of the industry during its necessarily limited life span: "we feel this way about it . . . Some day the ore will be gone. We might as well have a good time while it lasts, and get ourselves something to remember it by." High municipal taxes were also a source of direct public welfare in the Iron Range towns. When Hibbing mines closed, Victor Power hired more than a thousand workers—almost every able-bodied man in Hibbing—on the municipal payroll. Although reporters portrayed high municipal taxes as an example of wasteful public spending and shortsighted tax policies, the voices of Hibbing residents revealed that the high municipal tax system of the Iron Range cities in the early twentieth century was a coherent, populist response to the economic arrangements of extractive industry. The municipality was almost totally dependent on a single, fickle industry that would necessarily end at some point in the future. To Hibbing residents, the high taxes were a logical response to the financial realities of a mining town.[16]

In the long run, portrayals of the Iron Range's municipal tax system as wasteful were largely successful. High taxes on mines, even if used to finance public works and relief, were effectively painted as profligate and hindering the long-term growth of industry and municipal fiscal health. By implication, the underlying anticorporate philosophy behind the older tax system was vilified as well. A sense that the mining companies were unfairly taking profits that rightly belonged to the people of the Iron Range—illustrated by the residents' desire to "get ourselves something"—was consigned to an outdated and inefficient mind-set that needed to be updated for the mid-twentieth century. Thus, even apologists for high Iron Range taxes reveal how the fundamental mind-set about taxation had changed dramatically by the 1940s. Writing in response to the *American Mercury* article, one resident admitted that the Iron Range communities had wasted tax money in the past but argued that, "in recent years, the thought has spread that the mining companies and communities had much in common, that they prospered mutually." In this new atmosphere, Iron Range cities no longer passed large tax increases and now worried that tax rates made Mesabi ore too expensive in a competitive global market. In 1949, the city of Virginia's Chamber of Commerce officially announced its opposition to any increase in iron-ore taxation, citing the overall cost of iron ore and fears that higher taxes would lead to decreases in mining activity. By the middle decades of the twentieth century, attitudes about taxation on the Iron Range were in flux. Older ideas that high taxes were an appropriate means for communities to hold on to corporate profits slowly shifted toward a broad antitax sentiment that imagined taxes as a drag on productive industry. While these deep-seated changes simmered throughout the 1940s and 1950s, they moved to the forefront of political debate in the early 1960s.[17]

Economic Decline, 1955–61

The decade between the construction of the first taconite plants in 1955 and the passage of the taconite amendment in 1964 was a time of widespread economic anxiety across the Iron Range. Edward W. Davis's success in perfecting the taconite process led to the creation of two large taconite plants on the Iron Range by the mid-1950s. Reserve Mining and Erie Mining, both joint ventures by major steel firms, were churning out taconite pellets by the end of the decade. Yet the construction of the first two taconite plants did not guarantee that steelmakers would immediately turn to taconite as the predominant source of iron ore, nor did the success of the two Minnesota plants necessarily mean that future plants would be built in the Lake Superior region. Still a keen observer of the iron-ore industry, Davis recognized taconite's uncertain position in the late 1950s and early 1960s. He noted that taconite was still being mixed with natural ores in blast furnaces. More important, during the 1950s large deposits of rich natural ores were discovered around the world, including one in Canada's Labrador region that was richer than beneficiated taconite. "If taconite could not compete with foreign ores," Davis warned, "Reserve and Erie had made costly mistakes, and the future of Minnesota—indeed, of the whole Lake Superior ore-producing district—looked very dark."[18]

Dark clouds of economic uncertainty gathered across the Iron Range in these years. Throughout the late 1950s and early 1960s, the Iron Range's mining economy suffered a series of blows that led many to question whether mining would continue to exist in the region. While engineers such as Davis were debating taconite's merits, existing natural ore mines were closing down at an alarming rate. It is not surprising that these years were ripe for rearranging long-standing political alliances in the mining region. Once the iron mining industry's very existence was in question, political allegiances on the Iron Range were put in flux.

In the global iron-ore market, the superiority of natural ore from the Lake Superior region declined dramatically during the two decades following World War II. Minnesota's share of all U.S.

iron-ore purchases fell throughout the late 1940s and 1950s. While Minnesota was responsible for 68 percent of all U.S. iron-ore purchases in 1946, by 1957 it produced only 49 percent. At its highest production levels, iron ore from the Lake Superior region supplied nine out of every ten tons of ore used by the American iron and steel industry. By 1960, it was supplying less than half of the needed ore. These declines were a harsh illustration of the declining fortunes of natural ore mining on the Iron Range at midcentury. And it was not just the proportion of ore supplied by the Iron Range that was dropping. The total tonnage of natural ore mined on the Iron Range was falling sharply as well. Minnesota's production of iron ore dropped from 80 million tons in 1953 to 46 million tons in 1964. The overall context in which Iron Range residents confronted the taconite amendment in 1964, then, was amid a sharp decline in the natural ore mines.[19]

Underground mines in the Lake Superior region were particularly hard hit by the move toward taconite and a flood of imported ore. Older iron mining regions on the Vermilion Range, north of the Mesabi, and in northern Wisconsin and Michigan's Upper Peninsula relied heavily on underground mining to extract ore that ran deep below the surface. Costs to extract ore were higher in underground mines than in open-pit mines, which made the underground mines the first targets for shutdowns. The shutdown of an underground mine in Montreal, Wisconsin, was a particularly bad omen for underground mining in the region. The Montreal mine was the largest underground iron-ore mine in the Lake Superior region and when it was permanently closed in 1962 it still contained 7 million tons of unmined ore. The presence of so much ore in a mine that closed was an ominous sign that the underground mines were closing not solely because of depletion but because they were no longer cost-effective in comparison to taconite and foreign iron ore. The end of underground iron-ore mining signaled a new reality for Lake Superior natural ore: it was no longer the primary source of iron ore for the iron and steel industry. From this point forward, Lake Superior natural ores would compete with foreign sources and taconite to be used in the blast furnaces.[20]

The decline of natural ore mining, both in underground and in open-pit mines, had lasting effects on the number and types of jobs available in the region's mining industry. Throughout the 1950s and 1960s, the total number of workers engaged in iron-ore mining fell dramatically. Thousands of miners were laid off during this period and many faced no prospect of returning to mine work and the high salary it offered. Minnesota's iron mining industry shed more than six thousand jobs, some owing to seasonal layoff and others permanent cuts, in the last years of the 1950s. Yet, numbers of workers laid off do not tell the entire story of the era's decline. Even for those workers who retained mining jobs, hours and take-home pay were often cut as a result of the downturn. For example, workers at Oliver Mining were cut back to a four-day workweek in 1958. U.S. Steel later idled several mines entirely as the steel decline deepened. Whether permanently laid off or facing reduced hours, iron-ore miners in the late 1950s and early 1960s confronted a rapidly declining industry with few sure prospects for the future. This uncertainty about the future of iron-ore mining would play an important role in many miners' response to the taconite amendment.[21]

Beneath the dismal statistics, Iron Range residents experienced the decline in natural ore mining as a general feeling of malaise. Comments about work and life on the Iron Range at midcentury emphasized the region's declining fortunes and a feeling of despair that was omnipresent yet hard to locate. A national reporter described Hibbing in 1961: "The town is populated by 18,000 vigorous people . . . Once fabulously rich from ore taxes—the high school is an opulent reminder of those days—Hibbing has fallen on hard times. Many people are out of work and worried, and here one can see clearly the impact of Minnesota's growing economic problems." Another reporter wrote that the outlook for the Great Lakes iron-ore mines was "the bleakest in more than twenty years." Hard times on the Iron Range were evident both in mining statistics and evaluations of the area's future prospects.[22]

In many ways, overall quality of life was declining for Iron Range residents during the 1950s and 1960s. This is nowhere more evident that in the overall drop in the region's population during the immediate postwar era, which saw a steady outflow of residents to other parts of the state and nation. In 1964, an administrator from one iron mining county in Michigan described the exodus: "the fact is that everyone is leaving the area. We have always exported young people. Now we are exporting the middle-aged." Minnesota's mining towns faced similar population loss during these decades. The twentieth-century population of Eveleth, Minnesota, peaked in 1930 at 7,484 residents, but the city lost almost one-third of its population by 1970. The population of Virginia, Minnesota, stood at 14,034 in 1960 but had dropped to 11,056 by 1980. The decline in population was especially striking in the context of the nationwide population boom after World War II. The iron-ore mining regions were losing population at the same time that suburban schools and the Sunbelt region of the southern and western United States were rapidly gaining residents, illustrating both the uneven nature of the postwar economic expansion and the comparative poverty of the mining region. Population decline challenged the very existence of several small towns on Minnesota's Iron Range. In 1964, a consultant recommended that the towns of Virginia, Eveleth, Gilbert, and Mountain Iron—four of the largest towns on the eastern Mesabi Range—combine into a single entity rather than "drift into the future without positive corrective action." Without a radical response to the population decline—and corresponding decline in city revenue—the consultant warned, "the outlook is grim indeed" for the towns of the Mesabi Range.[23]

Declining population affected life in the mining region on a personal level, causing familiar community pillars like the church to fundamentally change in response to hard times. Pastors and elders of Hibbing's various churches convened a rare joint meeting in 1960 to discuss how they could respond to poor economic conditions in the area. "We are all faced with problems that may cause vacancies," one minister wrote, "the 1960 census showed that the population is down and so are some of the mines . . . far too many have moved away . . . No one church can say that nothing bothers us. We are all in this economic situation together." As the economic decline spread to the churches and synagogues, it was clear that few aspects of everyday life in the region were immune from industrial decline.[24]

The plight of the Lake Superior iron-ore mining region reached a national audience in the early 1960s when a rising star in the folk music scene used the Iron Range's economic despair as the inspiration for a song titled "North Country Blues." Hibbing native Robert Zimmerman, better known as Bob Dylan, left the Iron Range for Minneapolis in 1959, quickly making his way to New York City and national fame on the folk music circuit. In a song that evoked the classic folk tradition of Woody Guthrie, Dylan described the iron-ore mining towns as blighted by economic decline. The song described a town where "the cardboard filled windows and old men on the benches tell you now that the whole town is empty." Illustrating the historical context of his upbringing on the Iron Range in the 1950s, Dylan's lyrics connected despair in the iron country to foreign ore: "They say that your ore ain't worth digging, that it's much cheaper down in the South American towns where the miners work almost for nothing." Although Dylan said little about his Iron Range upbringing during his early career, the lyrics of "North Country Blues" gave voice to the region's harsh economic realities in the early 1960s.[25]

Beneath feelings of despair and decline were the hard fiscal realities facing Iron Range towns. With tax revenues declining because of population loss and a move away from natural ore, the towns confronted fiscal collapse. Using taconite as a source of iron ore proved especially problematic for the towns, because taconite's increased value as iron ore led to a corresponding devaluation of natural ore reserves and, thus, a declining overall tax base for many of the towns. For example, during 1957 hearings before Minnesota's Tax Department, mining companies protested the valuations of existing natural ore mines, arguing that these deposits were increasingly worthless in the taconite era and should be taxed accordingly. The Iron Range communities and school districts vigorously protested these devaluations, citing the financial harm that would befall their towns with the lowered tax base. Illustrated by census data, mining statistics, song lyrics, and general feelings of decline, the 1950s and early 1960s were a time of high anxiety in the Lake Superior iron-ore mining region. The older natural ore mines were obsolete as taconite and foreign ore upended the global steel industry. When Iron Range residents, business leaders, and state politicians confronted the taconite amendment and the call for lower taxes in the mid-1960s, a decade of economic uncertainty and industrial decline was a key factor causing many residents to abandon older, anticorporate beliefs and support tax breaks for industry.[26]

One reason for the Iron Range's decline was a flood of rich natural ore from abroad that entered the U.S. steel market in large quantities for the first time. Throughout the early twentieth century, iron ore from international mines occasionally made its way to American blast furnaces, especially those on the Eastern Seaboard or the South. For the bulk of American steelmakers located in the Great Lakes region, however, high transportation costs and easy access to the rich Lake Superior ores limited the use of foreign iron ore. This situation changed dramatically after World War II. As described in chapter 1, worries that Mesabi natural ore was dwindling caused mining firms to invest heavily in international iron-ore mines, leading to rich discoveries in South America, Canada, and Africa during the 1940s and 1950s. Several U.S. steel firms focused their investments on Canadian iron-ore deposits beginning in 1949. By the early 1960s, Canadian mines were exporting millions of tons of iron ore to U.S. steel mills. The development of a large iron-ore deposit on Canada's Labrador peninsula led industry observers to claim that Canada would soon overtake the United States as the largest exporter of iron ore in the world.[27]

South America was also a focus of development for the mining companies, where explorers in the 1940s found "a great iron mountain in the Venezuelan jungle." In 1950, U.S. Steel began exploiting a massive new discovery of high-quality Venezuelan iron ore. This new find was described as bigger and more pure than the Mesabi Range. There was a certain irony in the flood of foreign ore, developed in response to fears of depletion on the Mesabi and now threatening still-productive Mesabi mines. As a 1963 report put it: "what started out as an international developmental effort to supplement home ores now appears to threaten to supplant them." The report noted that in the 1950s and 1960s, imports of iron ore increased to 33.7 million tons, from an earlier 11 million tons, while domestic ore production declined. These mines were in operation by the mid-1950s and were soon competing with the Lake Superior ores.[28]

The influx of foreign iron ore dramatically changed relations among business, labor, and the mining communities on the Iron Range. International steel firms first sought foreign ore sources amid fears that Mesabi ores were quickly running out during World War II. After the war, however, the steel firms discovered that foreign ore had attractive advantages over domestic natural ore. What the steel industry found irresistible was the "higher iron content of the foreign ore . . . rock bottom labor costs, comparative freedom from taxes and government regulation, [and] a scale of operation that permits use of the latest and largest equipment and highly efficient ocean shipping arrangements." These advantages—lower labor costs, freedom from regulation, and the ability to build an enterprise from scratch at the optimum scale—defined the postwar wave of economic globalization that eventually displaced much of the older industrial economy in North America and Europe. Iron-ore mining was a forerunner to the later international capital flight.[29] As a forerunner, however, the mining firms struggled to implement a system of global industrial production. For instance, Hanna

Mining Company's attempt to build a low-cost iron-ore mine in Brazil was consistently thwarted by Brazilian politicians who were upset that the country's natural resources were being mined for export with little benefit to Brazil.[30] Nevertheless, the new regime of low-cost global production arrived in the iron-ore market by the first years after World War II, a development that forced American mining companies, labor unions, and local governments to revise older relationships in the face of new global competition.

In the Lake Superior mining region, the flood of foreign ore led industry leaders to call for a new spirit of cooperation among business, the unions, and government officials. Threats from foreign ore, executives argued, meant that older antagonisms—especially the vigorous anticorporate rhetoric of the unions—would lead to the decline of the entire industry. In a 1959 speech detailing the emerging foreign pressures to the iron-ore market, a steel industry expert argued that foreign competition demanded new cooperation among industry, government, and labor:

> The trend toward increased use of foreign ores and the possibility of removing part of our iron making facilities to foreign countries carry many implications . . . Even though we want our friends and neighbors to prosper, we must still maintain a strong and thriving iron and steel industry. To do this in the future is going to require much closer cooperation among government, management, and labor, than has been true in the past. We can no longer afford extensive strikes or lockouts, or excessive taxation.[31]

What is striking about these postwar calls for cooperation is that they were aimed not just at unions, but also at government. Facing the threat of foreign ore, business leaders turned to both labor and government to demand concessions.

Political debates over how much the taconite industry should be taxed did not take place in a vacuum. Behind debates in the halls of government was the hard reality of sharp economic decline on the Iron Range. For politicians, mining executives, and industrial workers, the prospect of long-term economic decline lurked beneath the specific tax policies. The grassroots pressure to lower corporate tax rates on the mining companies—pressure that contradicted a generation of anticorporate radicalism on the Iron Range—reveals a pragmatic core beneath the veneer of working-class radicalism in the early and mid-twentieth century. Many Iron Range miners may have been anticorporate crusaders, but the practicalities of a job and its paycheck ultimately trumped political ideology.

The 1961 Taconite Amendment Effort

Long-simmering debates about mining taxation led to a major contest over taconite taxes in 1961 when mining companies and Republican politicians first suggested an amendment to Minnesota's constitution to lower the tax rate on taconite producers. The immediate context for the 1961 amendment was growing concern that taconite plants would be built outside Minnesota. Taconite boosters in Minnesota argued that the state's high level of taxation on iron ore—the legacy of earlier attitudes toward the mining companies—now made the state "uncompetitive" relative to other states where taconite could be produced. For example, the IRRRB described New York's mining taxation situation as far more favorable than Minnesota's in 1945. At the time, New York was a main competitor to Minnesota for future taconite processing plants. According to one report, "The New York mining operations have a favorable tax situation. Mining is not subject to any special form of taxation whatever and the ordinary ad valorem tax burden is not heavy. Unquestionably the fear of being saddled with a high level of ad valorem taxes as our high grade ores become exhausted has tended to discourage heavy investments in Minnesota low grade ores."[32]

Among the most persuasive supporters of taconite tax incentives in Minnesota was Edward W. Davis, the mining engineer who created the taconite processing technology in his University of Minnesota laboratory. By the late 1950s, Davis was retired from the University of Minnesota and had moved to Silver Bay, Minnesota, where he worked as a consultant for the Reserve Mining Company. While the elderly engineer looked with pride on the Reserve Mining taconite plant, named the E. W. Davis Works in his honor, he expressed growing alarm at the course of taconite development after the Reserve plant was built in 1955. The first two taconite plants were built in northeastern Minnesota in the middle of the 1950s, but subsequent iron-ore pelletizing facilities were constructed in other parts of the United States, and even in Canada. Davis, who always assumed that his taconite research would benefit the Iron Range, was dismayed by the creation of an iron-ore pelletizing industry outside of northeastern Minnesota and argued that the Iron

Range tax system was pushing investment away from Minnesota. In 1958, he published a newspaper article arguing that the recent construction of iron-ore pelletizing plants in Canada, Michigan, and Wyoming—but not in Minnesota—was the result of a "business climate" that was unfavorable to investment by mining companies. He noted that the Iron Range communities depended on approximately $50 million in annual taxes from the mining industry, and as the mining industry switched to taconite and older natural ore mines closed, this tax source would dry up. Taconite operators, he wrote, believed the Iron Range communities would quickly shift this large tax burden onto the "big, juicy taconite plant that is anchored in place" and were therefore reluctant to invest in Minnesota. Passing a law about taconite taxation would not solve the problem, he argued, because taconite companies "know that politicians smart enough to enact a law are smart enough to modify it or repeal it." Davis was a trusted figure in Minnesota politics by the late 1950s and his championship of tax cuts for taconite helped to move the issue into the center of the state's political debates.[33]

Davis also took his appeal to the Iron Range communities that were suffering from the downturn in natural ore mining. He urged civic leaders to take direct action and lower their municipal taxes as an incentive to investment. In a meeting with Eveleth officials, Davis recommended what he would do to promote economic development in the region. He first told city officials that they needed to abandon old thoughts about tax revenue from industry and should focus exclusively on payroll taxes. Next, he told the Eveleth gathering, "If I were running this place I would send a committee down to the Oliver [Mining Company] office and ask them what could be done to get them to build a plant here. Give them the land? Rebate their taxes? Other communities do this to get industry. Why not this area[?] It seems perfectly plain that you must either get new industry or die."[34]

The idea of a constitutional amendment lowering taxes for taconite producers first came into focus in 1960. Pushed aggressively by politicians and business executives pursuing mixed objectives, the amendment moved to the center stage of state politics in the first year of the new decade. Outlines of a possible amendment first came from the mining companies and their advocates. In 1960, an attorney for a Duluth-based interest group connected to the mining industry proposed an amendment to Minnesota's constitution to limit taxes on taconite to the rate they would pay under Minnesota's income-tax law. Given the historically high rate of taxation on mines, the proposed amendment would cut taxes for taconite producers and, more important, guarantee that they would not be raised at a higher rate than the general state corporate income tax. The amendment was shaped for legislation in late 1960 and in the spring of 1961 it was proposed before state legislators. As introduced in the Minnesota House of Representatives, the 1961 amendment offered mining firms a guarantee that taconite operations would not be taxed at a higher rate than other manufacturing companies in Minnesota. In return, mining firms were required to build plants with a minimum capacity of 21 million tons of taconite production per year within eight years, and invest at least $250 million in new taconite plants. The 1961 amendment evolved as it worked its way through the Minnesota Senate and House of Representatives, but its essential elements remained intact: the people of Minnesota would constitutionally limit the amount that taconite could be taxed and, in return, the taconite industry promised jobs to a depressed region of the state. In other words, tax cuts equaled jobs. This equation proved a recurring mantra throughout the late twentieth century, on the Iron Range and in depressed industrial regions nationwide.[35]

The mining corporations that initially proposed the amendment were obviously strong supporters of the measure. In various public forums, mining company presidents and public relations officials hammered home the message that economic growth on the Iron Range depended on a guarantee of lower taxes. This message was sold through rhetoric that emphasized the primacy of jobs as tools for economic development. Speaking to journalists in 1960, the president of Oliver Mining Company argued, "a conviction that 'jobs are more important than taxes' is needed if state and local reliance on iron ore taxations are to be replaced by alternate sources of tax revenue." Turning to jobs and payroll taxes as the primary source of municipal revenue on the Iron Range would dramatically lower the tax obligations of the mining companies. Rather than taxation on iron ore—the historic source of funding for Iron Range towns—the new system of payroll taxation would make individual Iron Range workers and homeowners responsible for funding their government.[36]

Mining firms began complaining about high corporate taxation on taconite in the 1950s, arguing that their tax burdens were increasing dramatically while natural ore production was declining.

Many state legislators were initially skeptical of the mining companies' tax complaints, remembering that they had voiced similar qualms during the Great Depression. The experience of World War II confirmed some lawmakers' convictions that mining companies were "crying wolf" on the tax issue when mines abandoned during the Depression—ostensibly because of Minnesota's high taxes—were reopened during the war. Yet the influx of foreign iron ore during the 1950s dramatically changed the world market for ore and gave the mining companies far greater leverage in their threats to close down Iron Range mines. With proven reserves in foreign countries, international iron ore became "the mining companies' most potent weapon" in their push to lower taxes, allowing them to threaten to leave Minnesota altogether if taxes were not lowered. The rise of low-cost foreign mines created a double incentive for lowering taxes in Minnesota as the economics of the iron-ore market demanded lower production costs while foreign ore also made the threat of capital flight much more tangible.[37]

The mining companies certainly wanted to lower their tax obligations, but support for the amendment drew on less selfish motivations as well. Large taconite plants were hugely expensive. They required financing that was beyond the possibility of even the large steel firms, among the most capital-intensive industries in the world at the time. Steel firms mitigated the enormous capital requirements of the taconite plants partly by forming joint ventures to build plants, which would then split their taconite production among the different steel firms. Only U.S. Steel was large enough to finance a solely owned taconite plant. Additionally, by the early 1960s the steel companies no longer had access to government-backed funding that provided financing for the first taconite plants in the 1950s. The Erie Mining plant built in the mid-1950s was financed in part through a federal program intended to promote rapid industrialization, especially in the steel industry, in preparation for the Cold War. These federal loan programs were no longer available in the early 1960s. Taxes were an important component of financing taconite plants because the steel firms could only borrow the enormous sums necessary based on stable projected costs. If taxes could increase dramatically, banks and insurance corporations would not loan the money to the steel firms. As mining company executives told Minnesota lawmakers, they needed an amendment to convince outside investors that they could safely put money behind the taconite plants. The financial complexity behind corporate support for the amendment illustrates how many different motives were at work in the initial push for a taconite tax cut.[38]

Minnesota's Republican politicians had a more complicated relationship with mining taxes than the mining companies. Although Minnesota Republicans soon became vigorous supporters of a taconite amendment, their initial enthusiasm for the measure was driven by the DFL's political control on the Iron Range in the early postwar decades. By the early 1960s, the Iron Range was among the most reliably liberal regions of the state. In a 1960 campaign speech in Hibbing, for example, John F. Kennedy joked that the Iron Range was "the strongest Democratic area that I have seen in this campaign." In this unfriendly environment, Minnesota's Republicans were eager for any issue that might bring Iron Range voters into their fold or at the least shake their rock-solid DFL support.[39]

At its core, Republican support for the taconite amendment was driven by a desire to split Iron Range industrial workers from the leadership of the DFL and the labor unions. The idea of using taxation as a wedge issue on the Iron Range was floated by Minnesota conservatives in the late 1950s. As a campaign observer noted at the time, the Minnesota GOP hoped to use iron-ore taxation as "a sure-fire way to crack the traditional DFL stronghold of northeastern Minnesota."[40] The deepening economic crisis on the Iron Range in the early 1960s provided the necessary impetus to move the tax issue to the forefront of public debate. Minnesota's DFL politicians, drawing on New Deal rhetoric, emphasized strong labor unions and an active government presence—funded by high mining taxes—as the guarantors of prosperity in industrial regions such as the Iron Range. These promises increasingly rang hollow amid widespread shutdowns in the natural ore mines and the prospect of long-term job loss and economic decline. Liberal political rhetoric was built on the idea of constant economic growth—the rising tide that would lift all boats—and many liberals were at a loss when faced with the prospect of industrial decline in the mining region.

Conservative politicians in Minnesota realized that the economic collapse on the Iron Range opened space for political realignment. A tax cut for the taconite industry was a particularly useful wedge issue for the GOP because it simultaneously appealed to industrial workers by promising job creation while forcing DFL leaders to either renounce their earlier support of higher corporate taxes in the name of job growth or argue against job creation because of a largely ideological commitment to not lower taxes on corporations. The GOP's stance pinned DFL leaders between two difficult positions, each of which was sure to alienate some voters. When the DFL eventually opposed the 1961 amendment, the state Republican chairman argued that the DFL was apparently "against new jobs and opportunities on the Iron Range."[41] The amendment's political rhetoric counterposed jobs and taxes, suggesting that politicians had to choose between one or the other. What went unnoticed at the time was how this choice would have seemed ridiculous on the Iron Range only a few decades earlier when city officials pushed for high taxes precisely because they needed money for public jobs during hard times. The equation of low taxes and jobs reversed the prevailing ideology from the early-twentieth-century Iron Range.

The mechanics of the amendment also put DFL politicians in a bind. By pushing tax cuts via a constitutional amendment, which had to be voted on in a general election, the GOP added another layer of complexity to the political calculations. DFL opposition to the amendment could be framed not just as opposition to the tax cuts, but as opposition to democracy itself by refusing to put the issue before voters. Political observers at the time noted that GOP support for the amendment was fueled largely by political opportunism. Republican governor Elmer Andersen, in particular, pushed the amendment as a political wedge on the Iron Range. As one reporter put it, Governor Andersen turned the 1961 amendment into a successful "whipping horse" for votes on the usually liberal Iron Range. In a memorable Cold War phrase, a Minneapolis reporter described Andersen as "hanging onto [the amendment] like Yuri Gagarin," believing that "he has a political missile by the tail." Although the political opportunism of GOP support was clear, by framing the amendment as a stark choice between jobs and taxes Minnesota Republicans sharply constrained the options available for DFL leaders.[42]

Minnesota's DFL Party leaders thus faced a series of difficult decisions with regard to the taconite amendment. On one hand, the party represented its constituents in the heavily liberal Iron Range. Many Iron Range liberals intimately understood the region's economic plight and were eager to support any government plan that would alleviate the region's distress. On the other hand, one pillar of Minnesota's midcentury liberalism was close government oversight of corporations. From this perspective, the idea of cutting taxes on some of the world's largest corporations was politically anathema. Throughout the early 1960s, these competing positions roiled liberalism in Minnesota.

The 1961 taconite amendment opened up a sharp rift within the DFL's leadership that not only exposed differing ideas within the party, but pitted some of the state's highest-ranking DFL officials against one another. The rift was especially acute in Minnesota's House of Representatives, where the DFL's two leading legislators split over the amendment. The amendment's primary DFL supporter was liberal leader Fred Cina of Aurora, an Iron Range lawyer who put aside his antipathy for the mining companies to support job creation in the region. Opposition to the amendment was spearheaded by Donald Wozniak of St. Paul, thought to be the second-most powerful DFL member in the House after Cina. The amendment's ability to divide the two leading DFL representatives demonstrates how effective a wedge issue the GOP and mining companies had created with the proposed amendment.[43]

The taconite amendment forced liberals such as Wozniak to take a stand against the amendment based on an anticorporate philosophy that seemed out of touch in the face of economic decline on the Iron Range. Wozniak opposed the amendment because he believed it was a ploy by U.S. Steel to take advantage of Iron Range unemployment to lower its tax burden. He insisted that U.S. Steel would not build any taconite operations if the amendment passed. "This idea of providing jobs on the basis of a constitutional amendment is probably one of the most bitter hoaxes . . . ever conceived upon the people of the Range, those that are suffering from unemployment and misery by taking advantage of the situation by trying to get a tax break for the richest company in the U.S.—I think it's unforgivable." Speaking to the USWA in Hibbing, Wozniak also railed against Republican support for the amendment, arguing that it was a cover for Republican efforts to reward the rich through tax cuts: "The amendment will not do what its proponents say it will do; that is, guarantee more jobs." According to Wozniak, the taconite amendment would instead increase the tax burden on homes, steal future revenue from Minnesotans, increase taxes on other businesses, and hurt the overall business climate in the state. For Wozniak, the amendment was nothing but a wolf in sheep's clothing: an attempt by U.S. Steel and Minnesota

Republicans to avoid their tax obligations by promising jobs in an economically depressed region.[44]

What made Wozniak's argument so difficult was the hypothetical nature of the amendment. Because the amendment could only stimulate potential future job growth in the taconite industry, it did not particularly matter to many people if it was, in fact, a ploy by the mining companies or Republicans to lower taxes. These lower taxes would only be an issue if new taconite plants were built, in which case there would be new jobs to offset the tax losses. There was thus little credibility to Wozniak's argument that Minnesotans needed protection from the ploys of U.S. Steel or Elmer Andersen, because those ploys would only take effect when real jobs were created. Wozniak seemed to appreciate this problem and at times described more substantive concerns about a taconite amendment. Speaking on a radio talk show in 1961, he explained his opposition to the taconite amendment by arguing that taconite companies were already taxed less than other companies around the state. He was hesitant to put tax concessions into the constitution, he said, and believed that mining companies' decisions did not hinge on tax relief. His concern about adding specific tax language to the constitution was well founded—such technical language was traditionally a matter of statute and did not belong in the constitution—but he was again put in the position of arguing the theoretical point that current tax rates were likely already low enough that they did not hinder job growth. This technocratic argument rang hollow when compared to the far more urgent cry that lowered taxes would lead to more jobs.[45]

Wozniak was not alone in his opposition to the amendment in 1961. The DFL formally opposed the measure that year against the wishes of Cina, the leading DFL representative. Many party leaders were worried about the precedent of writing specific tax concessions into the constitution. The DFL chairman described the party's opposition to the amendment on the grounds that it was improper to write tax relief for a particular industry into the state constitution. He argued that this question went to heart of the relationship between corporations and democracy and said he was appalled that the mining companies were trying to subvert democracy by writing the tax laws into the constitution. A DFL state senator worried that the amendment was the opening salvo in a series of calls for business tax relief, using jobs as a carrot for unnecessary government concessions. Yet the DFL leadership likely realized that these technical arguments seemed insignificant in the face of economic decline on the Iron Range. As a newspaper editorial observed, liberals seemed to be more concerned with "putting 'big business' in its place" than alleviating unemployment and economic hardship on the Iron Range. On some level, DFL opposition to the taconite amendment was based on a fundamental belief that government should not bow to the wishes of a major corporation. Liberal DFL lawmakers accused U.S. Steel of using "its position as one of the most powerful corporations in America to induce the people of this state to put into the constitution a special provision for the taconite procedure." Cataloging the reasons for the DFL's official rejection of the amendment, they argued that "it is morally wrong to bow to U.S. Steel's pressure, money and emotional demands." But these moral concerns were offset by conflicting worries about how government could alleviate economic distress on the Iron Range. While railing against the mining companies, the DFL opposition paid surprisingly little attention to the reality of economic decline on the Iron Range. Many Iron Range residents likely concluded that liberals in the DFL either did not care about them or, more problematic in the long run, had few answers to the problem of long-term economic decline.[46]

DFL opposition to the 1961 amendment was not without grass-roots support. A strain of anticorporate ideology was evident among many Iron Range residents and these radicals were strongly supportive of the DFL's stand against the amendment. Among the amendment's strongest opponents were older miners. Many of the miners who personally remembered the violent 1916 strike saw efforts to pass the taconite amendment as little more than a power play by the "interests," revealing a continued reliance on an older vocabulary of labor radicalism that the DFL tapped into. The thinking behind this rhetoric poured out in impassioned letters to state politicians. One enraged Eveleth resident wrote to Cina urging him not to support the amendment. The author portrayed U.S. Steel as a rich, greedy corporation that had "already placed a quarter of a million people in this immediate area . . . in an economic straight-jacket" and "are now asking the legislature and the people of Minnesota, to support them in their brazen attempt to shackle even more drastically, many generations yet unborn." Civic leaders on the Iron Range were also wary of the amendment, but their opposition was based on concerns about future tax revenue for their towns. Municipal officials worried whether the amendment would create enough jobs to offset the loss in municipal tax revenue and if it would shift too much of the fiscal burden onto homeowners. At a meeting of the Iron Range League of Municipalities on March 26, 1961, concerned Iron Range residents voiced opposition to the proposed amendment. Several municipal leaders suggested that they would only support the amendment if taconite companies could guarantee that it would create several thousand new jobs. Iron Range residents and civic officials were deeply conflicted over the amendment, split between worries about its possible impact on taxes while acknowledging that something had to be done to help the region.[47]

For DFL representatives who supported the amendment, such as Fred Cina of Aurora, the amendment was a pragmatic response to economic conditions on the Iron Range. Many DFL supporters of the amendment were hesitant to cut taxes on the mining companies but did not see another option to promote economic development. Cina's support for a taconite amendment was unsure in the winter of 1961. In a letter to one Hibbing resident, he cited Reserve Mining's recent expansion as evidence that "the people that are presently in the taconite industry are extremely happy with the many tax concessions which have been given to them." He wrote that he needed to carefully consider the proposed amendment and would only support it if it would guarantee that Oliver Mining would build a taconite plant on the Iron Range. Cina ultimately did support the amendment, setting up a division within the DFL between the pragmatists who wanted to lower taxes to spur taconite investment and the idealists such as Wozniak who prized antitax principle.[48]

Like the DFL, organized labor was sharply divided over the 1961 amendment. Labor leaders were closely connected to the DFL leadership and pushed the major unions to reject the amendment. Only a few days after it was proposed, leaders of Minnesota's AFL–CIO expressed strong opposition to a constitutional amendment. One labor leader pointed out that current taconite taxes of five to ten cents per ton were insignificant compared to "the 40-cent tax on a carton of cigarettes or the 5-cent tax on a gallon of gasoline." A Duluth labor leader expressed his strong opposition by telling the Tax Committee to "stop this kind of nonsense." Ultimately, the state's labor leadership came out in common opposition to the amendment. At a 1961 state meeting the AFL–CIO distributed thousands of flyers opposing the amendment on the grounds that taconite already received favorable tax treatment in Minnesota. Labor leaders in Minnesota understood the amendment through the lens of hostile labor–capital relations and rejected the amendment as a giveaway to the mining corporations.[49]

What ultimately complicated DFL and labor opposition to the amendment was not the philosophical implications of their opposition but the amendment's widespread popular support. Throughout the state, average citizens were strongly supportive of a measure that would possibly improve northeastern Minnesota's economy at seemingly little cost. More than any other factor, deep public support for a tax cut on future taconite producers split the DFL and labor leadership from the broad mass of their constituents.

When newspapers surveyed Minnesota residents on the taconite tax issue, they found that a majority supported lowering taconite taxes if it could help the Iron Range's economy. A May 1961 poll by the *Minneapolis Tribune* found that 57 percent of survey respondents supported lowering taxes on taconite firms. When another survey contacted residents of northeastern Minnesota, it found an even higher level of support for taconite tax cuts. In the Iron Range region, approximately 67 percent of residents supported lower taxes for taconite firms. There was clearly widespread support for a taconite tax cut that promised economic development. Voters did not hesitate to tell their elected officials to support the amendment. In March 1961, Fred Cina received a petition from three hundred Iron Range "voters and citizens" who strongly urged him to support the amendment to bring investment capital to the area. Strong popular support for a taconite amendment put DFL and labor opposition in the position of opposing the amendment based on abstract principles rather than the wishes of residents.[50]

Widespread public support for a taconite amendment was evident at a dramatic public meeting of the Minnesota House Tax Committee in May 1961 that revealed just how impassioned many Iron Range residents were in their support for the amendment. The packed meeting also dramatically illustrated how the tax issue split DFL and labor leaders from their rank-and-file constituents, leaving them to argue against the amendment in the face of strong public support. On May 8, 1961, more than three hundred supporters of the taconite amendment jammed a Tax Committee hearing to speak in support of the amendment. Many of the supporters were unemployed miners who made the long drive to the state capitol in

St. Paul to show their support. Speaking before representatives at the meeting, the unemployed miners and other concerned citizens spoke out in favor of tax cuts for taconite firms. Here the full impact of a decade of economic decline was made clear as miners a generation removed from the anticorporate radicalism of the prewar Iron Range stood before politicians demanding a tax cut for the mining companies.[51]

What supporters of the amendment said was that jobs were ultimately more important to them than ideological purity in opposing the mining companies. Describing the attitudes of many Rangers toward the 1961 amendment, one speaker told legislators: "I have contacted an awful lot of working people and they seem to feel the same way that I do. They're a little bit leery, they don't want anything in it that's going to hurt them, by the same token they want work and they don't feel that this amendment will hurt them." Another miner and union member told the committee:

> I believe that this amendment is the best thing that could happen to the people of northeastern Minnesota . . . Now we've got an opportunity here to get $300,000,000 for the people of northeastern Minnesota and are we going to turn it down? . . . I've got a job . . . so I happen to be fortunate. But what about my friends, their friends, their sons and sons? We've got to do something for them people. This committee right here can do it. We've got nothing to lose and everything to gain.[52]

As this speaker illustrated, the amendment was a tremendously difficult political issue for the DFL because it was oriented toward future growth. While many liberals remained opposed in principle to tax cuts for a profitable resource-extraction industry, they could not argue that the tax cut would bring any immediate harm to the state. Instead, the worst the tax cut would do is limit future tax revenues. As the out-of-work miners who packed the Tax Committee meeting made clear, cutting future taxes cost nothing in the present and promised future gain.

Some proponents of the amendment threatened DFL opponents with political punishment if they failed to support the amendment. One Iron Range representative who did not support the amendment was told that if he "voted in a political vein only with the union vote in mind," he would "find that the average rank and file of union members will think differently come next election."[53]

The amendment also highlighted the distance between the union leadership and the rank and file. Although officials in the USWA and AFL–CIO were adamantly opposed to the amendment on the grounds that it was an unnecessary tax cut for the mining corporations, a significant number of the rank-and-file union membership supported the amendment because it might promote job growth. Demonstrating this tension, an employee of Oliver Mining Company testified that he had contacted two hundred members of the USWA and that the vast majority of them favored the amendment: "We know that our future in the Iron Range as far as miners are concerned lies in the processing of taconite. We know that . . . we have more to gain than we have to lose." Not surprisingly, mining company executives were quick to exploit the disagreement between workers and union leaders on the issue. In a letter sent to all Reserve Mining employees, President Robert Linney expressed shock that the union would oppose the taconite amendment and the expansion of taconite plants: "I don't believe [the union leadership's opposition] represents the real opinion of the Union, as distinguished from minority groups in the Union." As a divisive wedge issue, the amendment split the DFL leaders from their supporters on the Iron Range and the union leadership from their rank-and-file membership. The distance between these two positions—a principled opposition to cutting taxes and a pragmatic push for economic development—was the space needed for a subtle political realignment on the Iron Range.[54]

Within the context of this sharply divided political atmosphere, the 1961 amendment's legislative trajectory increased the partisan rancor surrounding the amendment and all but ensured that a taconite amendment of one kind or another would be a political issue in Minnesota in the coming years. The support of many rank-and-file miners for the amendment was tempered by testimony from mining executives who refused to publicly commit to building taconite facilities even if the amendment passed. Given that the entire rationale for the amendment was to induce taconite investment, executives' testimony was a sharp blow to amendment supporters. Speaking before the Minnesota Senate Tax Committee on April 10, 1961, Oliver Mining president Christian Beukema—arguably the most powerful iron-ore mining executive in the nation—told senators that Oliver Mining had no immediate plans to build a taconite plant on the Iron Range. Beukema repeated his claim in front of Minnesota House members, flatly telling the House Tax Committee that he could not guarantee that Oliver Mining would build a taconite plant if the amendment passed. Even more damning was his unwillingness to promise that a taconite plant would create new jobs on the Iron Range. Pressed by lawmakers to describe the number and types of jobs that might be created in a future taconite plant, Beukema refused to estimate how many employees would be needed at a future plant. All he could promise was that Oliver Mining "would do all it could" to hire local employees.[55] What Beukema and other mining executives likely understood was that taconite processing was highly automated and technical work. The amount and type of labor required was much different from natural ore mines. This meant that a given quantity of taconite ore required far less human labor—although enormously greater quantities of machine labor—than natural ore and likely would require workers with highly specialized skills, perhaps meaning that workers would have to be imported from other regions.

Beukema's pessimism swayed several Iron Range lawmakers who were wavering over whether or not to support the amendment. For legislators who were initially skeptical of the mining corporations—historically, the default stance of Iron Range representatives—the statements confirmed that the mining companies were not serious about using the amendment to create jobs on the Iron Range. Several lawmakers withdrew their support for the amendment, claiming that the mining companies were not negotiating in good faith. One disillusioned representative, Jack Fena of Hibbing, proposed an alternative amendment to lend state money for taconite expansion. Rather than lowering taxes, Fena argued that direct loans would "put an end to the argument of whether our business climate is good or bad—an argument that no one can win and that serves only to divide the people of the Iron Range and the state."[56] The plan for direct loans was never adopted, but its existence signaled that there was dissatisfaction with the amendment even among those who favored government support for taconite.

Despite widespread public support for the amendment, liberal opposition in the Minnesota House Tax Committee ultimately killed the measure. Led by Wozniak, the House Tax Committee refused to take action on the proposed amendment, which effectively ended the possibility that it would go before voters. Governor Andersen intervened at the last minute to plead for the amendment, but his pleas did not prod the Tax Committee into action.[57] For supporters of the amendment, the manner in which it failed was particularly galling. The amendment was killed by a legislative maneuver in a House committee, preventing the full Minnesota House of Representatives from debating it. Additionally, by coordinating the amendment's demise in committee, DFL lawmakers were painted as opposing democracy because the amendment would have been voted on by all Minnesota voters.

Almost immediately after the amendment was killed in the Tax Committee, groups throughout the state cried foul. Officials from the Minnesota Junior Chamber of Commerce argued that DFL representatives "took away the inherent right of the people to vote" by refusing to put the amendment before voters. The group began a statewide petition urging politicians to reconsider the amendment and put the issue before the voters. Overall, the outcome of the amendment in 1961 tarnished the DFL's reputation as the party of industrial workers. Many Iron Range residents reconsidered their support for the DFL. As one letter put it, "I cannot but feel that the D-F-L Party—which I have always supported in the past—will bear the onus for not saving our iron mining economy." Governor Andersen immediately attacked the DFL for the amendment's failure. Speaking to an Eveleth radio station, he accused DFL politicians of "playing politics with jobs" and urged Iron Range voters to contact their representatives and urge them to support the amendment.[58] Among the sharpest criticism of the DFL's handling of the issue was an October editorial in the *Minneapolis Tribune* that accused the DFL of voting against the economy of northeastern Minnesota on the basis of a theoretical concern with taxation and big business. According to the editorial, the DFL was insulting the people of Minnesota by arguing that widespread public support for the amendment was the result of being duped by U.S. Steel. Instead, the *Tribune* argued that the GOP had adopted the liberal position on this issue by supporting the right of Minnesotans to vote on the amendment. Perhaps realizing just how politically unpopular their opposition to the amendment had become, DFL representatives quickly passed a resolution expressing support for taconite. The resolution was seen as ineffective, however, because it had no binding effect on future legislatures.[59]

In the aftermath of the amendment's failure in 1961, the political implications of tax-cut policy were clearly visible. By supporting tax cuts as a tool for creating jobs via taconite investment, the mining companies, Minnesota conservatives, and supportive DFL politicians split a wide swath of Minnesota voters from DFL and union leadership. In contrast, opposition from the DFL and labor leadership was based on abstract principles about corporate power and the relationship between the state and the mining companies. By holding fast to a principled commitment not to cut taxes, liberals opened up a new type of critique from Republicans who now argued that they, not the DFL, best represented the needs of industrial workers.

The 1964 Amendment

Failure of the amendment drive in 1961 did not end discussion of a tax break for taconite producers. If anything, arguments over the taconite amendment became even more heated after the amendment's initial failure. The ongoing rancor surrounding taconite taxation ensured that the issue would remain at the forefront of state politics in the early 1960s. During elections in 1962 the amendment was a major issue for Republican and DFL candidates, with Republicans arguing that the DFL was opposed to jobs on the depressed Iron Range. In contrast, the DFL argued that Republicans, including Governor Elmer Andersen, were a "handmaiden" to the steel industry. By 1962, the taconite amendment had become a popular wedge issue around which the state's politics began to pivot.[60]

While the taconite amendment was polarizing state politics, DFL supporters of the amendment worked to craft a measure that would gain the support of both parties. In early 1962, Cina introduced a new bill—but not a constitutional amendment—in the state House of Representatives that guaranteed that taconite would not be taxed at a higher rate than other Minnesota industries for a period of twenty-five years. Governor Andersen hailed the bill as a "major breakthrough," although he still believed a constitutional amendment would offer a stronger long-term guarantee to taconite firms. More important, other liberal DFL lawmakers supported this revised bill, an important consideration because their opposition had killed the 1961 amendment. Partisan wrangling continued, however, as the Minnesota GOP immediately pounced on Cina's revised bill as an ineffective half measure that would not create a strong incentive for taconite investment. GOP leaders pointed out that promises could not be enforced on future legislatures and wondered why liberals would not "go all the way" to support taconite on the Iron Range. GOP strategy by 1962 was to continue scoring political points by attacking DFL opposition to the amendment, so there was little incentive to support a compromise measure. Republicans then escalated the rhetoric surrounding taconite taxes, with one state GOP leader describing the DFL's actions as "callously stabbing the economy of Minnesota and its people in the back." From the Republicans' perspective, the amendment had become a valuable wedge issue that forced liberals into a difficult position. With the majority of voters supporting their position, conservatives felt little incentive to compromise.[61]

Indeed, those members of the DFL and organized labor who opposed the amendment faced an increasingly hostile public that was not interested in principled arguments against the tax amendment. With mounting pressure in favor of the amendment, DFL and labor leaders alienated themselves further from their rank-and-file supporters. Organized labor was unmoved by the groundswell of support for an amendment within its own ranks and leaders of the state's major unions dug in their heels in opposition. In 1962, the USWA, District 33, approved several resolutions expressing unified opposition to the taconite amendment and urging its members not to support any legislators who backed the amendment. Observers in the press were baffled by the union's rejection of the taconite amendment, calling it "one of the big puzzles" in state politics. Citing declining employment in mining and excitement about taconite, reporters expected the steelworkers' union to be a strong supporter of the measure. From a national perspective, however, the union's opposition was not unusual. During the second half of the twentieth century, labor unions throughout the United States struggled to respond to the concerns of members facing economic decline. In eastern Pennsylvania's anthracite coal-mining region, for instance, the powerful United Mine Workers of America offered little support to miners confronting the shutdown of anthracite mines in the 1950s and 1960s. Rather than work on behalf of its members, the union descended into widespread corruption and even murder in the face of mounting job losses. Historians Thomas Dublin and Walter Licht describe the union's response to industrial decline as "dismally inadequate." From a long-term perspective, the difficulty many unions had in responding to economic decline suggests that industrial growth was central to American unionism in the twentieth century. Unions had very few answers for workers facing long-term economic decline and, for many industrial workers, conservative promises to create jobs based on tax cuts were as viable as union alternatives.[62]

While the steelworkers' union dug in its heels, a few opponents of the amendment offered substantive critiques of the proposal. But their opposition rested on complicated and abstract questions about the appropriate place for tax policy within state government. One example of such detailed criticism came from the Minnesota Emergency Conservation Committee, which released a statement urging rejection of any future amendment. The committee argued that amendments were extremely difficult to repeal and that, if passed, "in twenty-five years . . . we might find ourselves faced with an amendment, which we wish had never been adopted." Arguments about the appropriate venue for tax legislation fell on deaf ears among the public at large, and polls revealed that a growing percentage of Minnesotans supported the amendment. A poll in early 1962 found that 70 percent of respondents were now in favor of a taconite amendment. Two-thirds of labor union members felt that the taconite industry was "very important" for the state's future economy. Faced with poll results showing broad support for an amendment, some DFL legislators argued that the public was simply wrong on the issue. Liberal Donald Wozniak insisted that the poll only revealed that "a great number of people are still confused" about the amendment. Illustrating the growing public disillusionment with DFL and labor opposition, the major Minnesota newspapers began to sharply criticize the DFL for its stubborn opposition. In an April 1962 editorial, the *Minneapolis Tribune* critiqued the DFL for opposing a taconite amendment on the principle that the steel companies should not receive tax cuts. Worse, by trying to "enforce its theoretical argument," the DFL was willing to "lose future taconite plants," the newspaper argued. Just as state Republicans had hoped, the taconite taxation issue had grown into a deeply divisive wedge issue that forced the DFL and organized labor to stand in opposition to a broadly popular measure. The taconite tax issue transformed liberal anticorporate principles from their earlier essence—populist support for average workers—into a stubborn refusal to do what was necessary in the face of industrial decline.[63]

The shape of the political debate over the amendment was upended in early 1963 when the USWA reversed its opposition during a secret meeting with the steel companies in Pittsburgh. In late 1962, Oliver Mining laid off 1,250 miners, fueling rumors that a permanent shutdown of the Iron Range's largest mines was imminent. Desperate, union leaders met with U.S. Steel executives in

CREATING A TAX-CUT CONSENSUS

Pittsburgh, bluntly asking what they could do to get the jobs back. Christian Beukema, head of Oliver Mining, told union leaders that only union support for a taconite amendment would offer hope for returning to work. Many union leaders had sharply criticized the amendment and found this condition a bitter pill to swallow. But the union ultimately supported the plan, admitting that it had few other options. Iron Range legislator Fred Cina confirmed that the miners were driven to support the amendment out of desperation rather than principle. The Iron Range miners were "desperate," he said, "they seek something . . . and they don't care what it is." Many Minnesota lawmakers were pleased by the surprise announcement. Labor opposition to the amendment had been a critical stumbling block in the failure of the 1961 amendment and labor's support virtually guaranteed that an amendment would pass. Indeed, within several hours of the announcement, the Minnesota House of Representatives generated several bills for a possible amendment. Iron Range civic leaders were also enthusiastic about labor's reversal. Hibbing's mayor, for example, said the agreement would hopefully be "the break we have been waiting for." Labor's switch, born out of desperation and fears of shutdown, signaled that principled opposition to tax cuts no longer appeared reasonable in the face of continued economic decline.[64]

Labor's reversal put the DFL in an even more difficult position. Minnesota liberals could stand behind their principled opposition to the amendment, at the risk of looking even more out of touch with economic misery on the Iron Range, or they could take a cue from labor and switch to support the amendment. At first, DFL leaders chose the former course of action, as they remained wary of the amendment and the labor–steel agreement. One DFL leader told reporters that the party would not be swayed "by a Pittsburgh pact." Another claimed that the union was pressured by U.S. Steel to make a bad deal and "conceded far more than they had to and far more than industry would have accepted." Although the DFL's argument that it could out-negotiate the steel industry where labor had caved was meant as a defense of workers, it was heard by many as an example of the DFL's increasing condescension toward rank-and-file union members. Leading the continuing DFL opposition was Wozniak. Even after the USWA came out in support of the amendment, Wozniak was still rallying the union to reject the taconite amendment as a giveaway to the steel companies. Invoking traditional populist rhetoric, he told the miners: "You know everyone talks about a free enterprise system and certainly I'm for it, but I am beginning to wonder what's so free about it. I am beginning to wonder how much control has the average worker, the average primary producer, the small businessman, got over the means by which he lives." Although many Iron Range residents likely agreed with Wozniak's philosophy, they now supported tax cuts as a practical measure in the hopes of creating jobs amid economic decline. The union's reversal left the DFL as the only major group opposing the amendment. Republicans were only too happy to take advantage of the continuing DFL opposition to the widely popular amendment. Throughout 1963, the GOP leadership hammered the DFL for its ongoing opposition, with a GOP leader emphasizing that the DFL was now out of step with "the steel workers union, the steel companies, the legislature, and the people of Minnesota."[65]

The second major development that reshaped the amendment debate in 1963 was a commitment by the steel companies, especially U.S. Steel, to build taconite plants if the amendment succeeded. For years, the steel companies had pointed to high taxes as an impediment to investment in taconite facilities. But in 1963 the major firms indicated that they were ready to begin building plants, provided that the amendment passed. U.S. Steel began planning a large taconite processing plant at Mountain Iron, but insisted that it was contingent on the passage of an amendment guaranteeing "fair taxes" on taconite for decades into the future. An executive with U.S. Steel was quoted as being "confident that the people of Minnesota will support the legislators' endorsement of the taconite fair tax amendment." The taconite plants were enormous technical facilities that required years of planning before construction could begin. Thus, U.S. Steel began designing a plant capable of producing 4 million tons of taconite per year prior to the amendment. Actual construction would remain contingent on the passage of the amendment. The steel companies' actions in 1963 highlight the complicated game of chicken played between the companies and the state on the tax issue. The closer the steel companies were to building taconite plants, the more Minnesota's citizens could see the tangible benefits of an amendment. Conversely, the more steel companies finalized their plans before an amendment passed, the more they called into question whether or not they absolutely had to have tax relief to invest in taconite.[66]

71

The combined weight of labor's support and the commitment of steel companies to build taconite plants pushed most DFL legislators to reverse their earlier opposition and support the amendment. Fred Cina again led the DFL support for the amendment. He proposed a compromise bill that limited the effects of tax equalization to a period of twenty-five years. The quarter-century limit pacified former opposition groups in the DFL—or provided a convenient excuse for them to switch their vote—and Cina's measure was widely supported by interest groups statewide. On March 18, 1963, Governor Andersen signed a statute into law that guaranteed Minnesota taconite producers "tax equity" with other manufacturing firms in the state. Although the statute was hailed as an important first step, legislators worked quickly to prepare a new amendment for the statewide ballot in 1964. By now, the amendment faced little opposition from either side. The amendment sailed through the Minnesota House of Representatives and the state senate quickly approved a bill putting an amendment before voters in 1964. Only nine liberal senators voted against it, voicing isolated arguments that the state was "giving away our resources to U.S. Steel" and suggesting that the state had caved in to the demands of the steel industry. Reflecting on the continued opposition of a few liberals, one conservative senator said their opposition would be "ludicrous if it weren't so sad." Many liberals were not enthusiastic supporters of the amendment but ultimately capitulated to the arguments put forward by the steel industry that economic growth on the Iron Range demanded the amendment. As one Iron Range lawmaker dejectedly claimed, "if this is the kind of thing the industry feels it needs . . . to invest millions of dollars, I think they ought to have it." The amendment ultimately approved to go before voters in 1964 guaranteed Minnesota taconite producers that their taxes—occupation, royalty, and excise—would not exceed the greater of then-existing ore tax levels or the ratio of taxes they paid relative to other state manufacturing industries. In essence, the amendment assured taconite companies that their state taxes could not rise disproportionately higher than those of other manufacturing companies in the state.[67]

In sharp contrast to the failed 1961 amendment drive, the 1964 effort received the support of organized labor throughout Minnesota. USWA president David J. McDonald made a special appearance at a Minnesota labor meeting in 1963 to announce his support for the taconite amendment. He told steelworkers that he supported the amendment "flatly and fully" with "no doubts or reservations." Although he acknowledged that many union members were suspicious of the steel companies' motives, he invoked the new labor–capital consensus by telling union workers to overcome their suspicion of the companies because "times have changed and so have attitudes." Union leaders made steady rounds of speeches and printed brochures urging their members to support the amendment. In a 1964 pamphlet for members, the USWA explained that the amendment would benefit all of Minnesota by offering jobs rather than relief dollars to unemployed miners: "The members of the United Steelworkers of America on the Iron Ranges of Minnesota want to be tax-payers and not relief recipients," the pamphlet urged. "Passage of the Taconite Amendment . . . will give these men jobs and at the same time save you the burden of costly relief through increased taxes." Speaking to fellow steelworkers, the union emphasized that the amendment was necessary to bring taconite plants to Minnesota rather than to other states or foreign countries. The USWA's local director urged union members to support the amendment in a speech in Duluth. Noting that taconite plants were being built throughout the United States but not in Minnesota, he argued:

> The United Steelworkers of America . . . feel that this trend
> will have to be stopped or northeastern Minnesota will be
> depressed for many years to come. Our only solution lies in
> the development of taconite plants, and here we find that the
> stockholders of certain steel companies are reluctant to invest
> their money in taconite plants in Minnesota unless they are
> assured of fair tax treatment for at least twenty-five years be
> cause of the tremendous investment involved.

This speech, in which a union leader was urging members to support a tax cut for the steel firms in the name of investment, reveals just how far the prospect of economic decline had shifted older patterns of Iron Range radicalism by the mid-1960s.[68]

Despite official union support for the amendment, however, many Iron Range residents were very ambivalent about passing the amendment. Many supported it simply because they wanted to do something—anything—that might promote industry on the beleaguered Iron Range. The sense that the amendment was a last chance was evident in the words of a Hibbing union official in 1964:

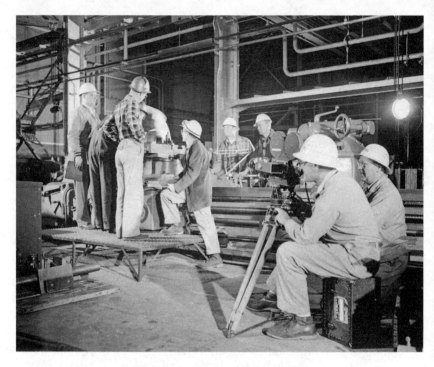

Taconite production was heavily promoted to Minnesota's voters and taxpayers during the 1950s and 1960s. This photograph shows a crew filming a promotional movie inside a taconite plant. Photograph by Duane Lundquist, circa 1958. Minnesota Historical Society Collection, N2.3 p149.

"I tell you something—this taconite amendment, by God, it's our last hope . . . If people vote the amendment down, we are finished here." Or, as an elderly Iron Range resident put it in a letter to a newspaper: "And that taconite question? Vote for it, people need jobs. Some say the companies don't pay enough taxes. What difference does that make?" The 1964 amendment had widespread public support on the Iron Range, but that support was fueled by desperate hopes for jobs and economic development rather than principled belief in tax incentive policy. The anticapitalist radicalism of the prewar Iron Range was jettisoned quickly when the mining industry was faced with long-term decline and job losses.[69]

After approval by Minnesota legislators, the amendment still required passage by a majority of Minnesota voters to become law. To ensure that the amendment passed at the ballot box, a powerful interest group was established to coordinate statewide publicity in

Donald Fraser (left) *and Walter Mondale were among the many prominent Minnesota politicians who supported the 1964 Taconite Amendment. Minnesota Historical Society Collection, Walter Mondale Papers.*

favor of the amendment. The Citizens' Committee for the Taconite Amendment was an early example of the single-issue political pressure groups that proliferated in American life by the late twentieth century. These small, powerful, and well-funded committees acted as though they were a grassroots movement by ordinary citizens while, behind the scenes, they were controlled by powerful lobbying interests.[70]

Throughout 1964, the Citizens' Committee worked feverishly to ensure that the amendment would pass. A statue honoring taconite was built in Silver Bay, Minnesota; the governor declared a "taconite amendment week" during a Minnesota Vikings football game; and the governor's wife promoted a "tell ten people" campaign of word-of-mouth support for the amendment. One day in October was declared "Education Day" and schools around the state were shipped taconite pellets and literature emphasizing the "industrial importance [of mining] to education in the state." Although

A sign in Chisholm, Minnesota, supporting the 1964 Taconite Amendment. The amendment benefited from a statewide publicity blitz that included signs, sporting events, mailings, and broadcast media. Photograph by Powell Krueger. Minnesota Historical Society Collection, 07025–3.

teachers were urged to present both sides of the taconite amendment debate in their classrooms, many apparently felt there was no reason to present the opposition views and just gave students promotional literature supporting the amendment. Iron Range officials held a "knighting ceremony" for the mayors of Minneapolis and St. Paul, along with the University of Minnesota's president.[71] One thousand Iron Range citizens arrived by caravan at a Minnesota Twins baseball game. At the game, they held a "home-plate ceremony" to give honorary hard hats to state officials. To cap off the event, a taconite-themed parade float carrying the "Taconite queens" was on hand. By election day, observers called the pro-amendment publicity campaign "the most massive ever mounted . . . on an amendment issue." Most of these events were little more than publicity stunts, but they had the desired effect. In case there was any doubt of support for the amendment, newspapers and DFL politicians actively campaigned for its passage. Most of the state's prominent politicians came out in support, including DFL politicians who had earlier opposed an amendment, including Hubert H. Humphrey, Don Fraser, Karl Rolvaag, Eugene McCarthy, and others. The *Minneapolis Tribune* not only endorsed the amendment, it also ran a five-part series explaining exactly how the amendment would lead to a taconite industry and revitalize the Iron Range's economy. The articles suggested that the entire fate of the Iron Range hinged on the amendment: "The future of iron mining in Minnesota is in that rock [taconite]," the articles explained. The well-funded and well-organized bonanza of publicity in favor of the amendment virtually guaranteed its support by voters in the November general election.[72]

A small number of liberal legislators did form an advocacy group to oppose the amendment. The group, called the Constitutional Protection Committee Inc., was worried about the amendment's implications for the state constitution. As one member put it, "it's the principle of the thing that we object to as much as the taconite amendment." By this time the antiamendment liberals were reduced to arguing over abstractions of constitutional principle. The group soon admitted that it did not have the financial resources to match the pro-amendment Citizens' Committee for the Taconite Amendment. Whereas the Citizens' Committee had gained more than five hundred endorsements, the antiamendment group had a difficult time even finding audiences willing to hear its side of the argument. As amendment supporter Cina put it during a public debate on the amendment, abstract worries about Minnesota's constitution were meaningless in the face of deepening Iron Range depression. "With men going hungry on the Iron Range because they can't get jobs, words about the sacredness of the constitution mean nothing," he told the audience.[73]

The publicity campaign ultimately paid off. On November 3, 1964, Minnesota's voters approved the taconite amendment by a wide margin and tax equalization for taconite entered the state constitution. Newspapers hailed the amendment's passage as a wholehearted signal to the steel companies that Minnesota welcomed investment and knew that low taxes were crucial to the new industrial economy. The *Minneapolis Tribune* declared the amendment "a great proclamation to the nation's steel industry that Minnesota wants jobs and investments and is willing to assure equitable taxation in return." A St. Paul newspaper concurred: "The resounding victory opens the way to an era of expanding industry and increased prosperity on the Iron Range, with accompanying benefits for the entire state."[74]

An ore boat loads taconite pellets hauled by rail from one of the new taconite mills built on the Iron Range during the 1950s and 1960s. Photograph by Gordon Ray. Minnesota Historical Society Collection, I.198.46.

The steel companies lived up to their promises, too. Within twenty-four hours of the amendment's passage, U.S. Steel publicly announced that it would begin construction of a huge taconite plant at Mountain Iron. The company projected that the plant would require four thousand construction workers and take three years to build. Hanna Mining and Jones and Laughlin Steel Corporation also announced plans to build taconite plants. Although these announcements certainly heartened the amendment's supporters, the instant announcements again raised the thorny question of whether the steel companies would have built taconite plants without the tax incentives granted by the amendment. In other words, was the amendment absolutely necessary for corporate investment or had the steel companies used it as leverage to lower their tax rates? Around the world, steel companies invested approximately $1.5 billion in iron-ore pelletizing plants by 1965. Almost $1 billion of that total was invested on the Mesabi Range. This huge corpo-

Two workers watch taconite being dumped from a railcar into a crusher at a taconite mill in the mid-1960s. Taconite mills were highly automated and utilized novel technologies, such as the machine pictured here that tipped railcars without uncoupling them. Photograph by Basgen Photography, circa 1965. Minnesota Historical Society Collection, HD3.24 p11.

rate investment certainly rejuvenated the Iron Range economy throughout the late 1960s and early 1970s, but it did little to halt the rapidly growing international competition in the iron-ore market. International expansion of iron-ore pelletizing plants was even more spectacular than the Mesabi Range boom. Industry experts expected pelletizing capacity in Africa, Asia, and the Pacific Rim to grow twentyfold from 1965 to 1975. The taconite amendment addressed the temporary malaise of economic depression on the Iron Range, but it did little to halt the underlying disease: the growth of an increasingly globalized economy in the late twentieth century.[75]

The long-term decline of the Iron Range mining economy was not the foremost concern of most Iron Range residents immediately after the amendment's passage. Instead, residents experienced the amendment and the subsequent taconite building boom in deeply personal terms. According to one national reporter, the amendment's real benefit to Iron Range residents was a "psychological boom" resulting from the belief that better times were ahead. Observers described a new economic climate of "unrestricted optimism" on the Iron Range after the 1964 taconite amendment passed. One older Mountain Iron resident's story reveals how the political maneuverings of the taconite amendment were experienced in deeply personal terms. As a fifty-seven-year-old man told a reporter, the amendment's passage gave him hope that he could spend his later years surrounded by family: "I had two sons and a daughter who left town because there were no jobs. Now, they may come back." For Iron Range residents, the amendment did just what proponents had promised: by cutting taxes, new jobs flooded into the region and the depression of the 1950s and early 1960s was soon over.[76]

Amid the construction boom—which included new schools as well as new taconite plants—the region paid little attention to the gradual shift away from anticorporate rhetoric and toward a new emphasis on labor, capital, and government consensus in the name of low taxes and job growth. As one reporter wrote, the taconite amendment "effectively curbs eighty years of strife, suspicion, and—at times—bloodshed between Minnesota, the miners, and the mining companies." The amendment thus came to be seen as the turning point from a pre-1964 history of antagonism between labor and capital and a post-1964 story of consensus aimed at job growth and low taxes. A student essay from 1963 poignantly demonstrated how individual residents, especially young people on the Iron Range, internalized the new era of consensus as a sweeping away of the old habits of conflict between labor and capital. As Gayle Anderson, a senior at Roosevelt High School in Virginia, Minnesota, and the winner of a 1963 essay contest sponsored by Oliver Mining Company, wrote: "This generation must shed itself of the inbred inertia, the deluded mentality which would cling to the past, the suspicious mind which regards every action of the mining companies with bitter distrust." Anderson's evocative call for harmony between miners and the mining companies is striking in the context of the Iron Range's long history of labor conflict. Young Iron Range residents such as Anderson were only one generation removed from the radicalism of Victor Power and the belief that the mining companies were stealing ore that rightfully belonged to the people of the Iron Range, with high taxes a partial recompense for their loss. To Anderson's generation, taxes were a necessary evil that had to be carefully monitored lest they prevent corporate investment and job growth.[77]

Conclusion

What were the long-term consequences of the 1964 amendment for the Iron Range? From this longer historical perspective, it is clear that the amendment was not a magic bullet capable of fully solving the Iron Range's unemployment problem. Taconite mining and processing were highly mechanized, meaning that fewer workers were needed than in natural ore mining. They also required highly skilled employees or workers with different skills than those developed in natural ore mining, meaning that many unemployed natural ore miners could not move easily into the taconite plants. Within a year of the amendment's passage, business analysts admitted that taconite production would never revive mining employment on the Iron Range to its previous levels. Experts predicted thirteen thousand jobs in taconite by 1975, still below the twenty thousand miners employed in the Lake Superior mining district in 1957.[78] Although taconite promised new jobs on the depressed Iron Range, it never matched the number of mining jobs available before the downturn of the 1950s. Historians of mining regions and resource-extraction industries often describe the "boom-and-bust" cycle of one-industry towns relying on volatile industries such as iron ore.[79] Focus on the cyclical employment swings in the mining industry, however, can obscure the overall downward trend in mining employment on the Iron Range during the second half of the twentieth century. The amendment did create a taconite boom, but increased employment in taconite mines only slowed the rate of mining's decline on the Iron Range.

Increases in taconite mining also fueled the full-fledged abandonment of the older natural ore mines on the Mesabi and nearby iron ranges. Expansion of taconite processing on the Mesabi Range caused the abandonment of the smaller Cuyuna Range to the west of the Mesabi. In an area that had once supported twenty-eight iron-ore mines, the last underground mine in Minnesota closed in 1967. According to one mining expert, the iron ore remaining in the Cuyuna Range "was not well suited to present separation methods." Most of the underground iron-ore mines in the Lake Superior district were shuttered by the mid-1960s because they could no longer compete with the open-pit mines producing low-grade ores for pelletizing. In a desperate effort to create a taconite industry, few considered the long-term consequences that tax policies might have on the rest of the mining industry.[80]

The 1964 taconite amendment—including both its contested political history and its ultimate passage—also suggests important lessons for the history of tax policy in the twentieth century. The reversal of tax sentiment on the Iron Range demonstrates how a tax-cut consensus emerged in the postwar decades and how the political discourse of low taxes was capable of turning even the most anticorporate Americans into ardent proponents of tax cuts for corporations in the name of job growth. Democratic resistance to taxation has a long and venerable history in the United States. Grassroots support for the taconite amendment is one example of what political historian Julian Zelizer describes as the long history of "democratic pressure from voters to maintain low rates of taxation."[81] Yet ardent support for high taxation on mining companies on the pre–World War II Iron Range—best exemplified by Hibbing mayor Victor Power—suggests that industrial workers' attitudes toward taxation were far more complicated than a general and unchanging hostility to taxes. The early-twentieth-century Iron Range was pro-tax, provided that the mining corporations, rather than the miners, were paying those high taxes. While someone else was paying, the miners gladly, sometimes gleefully, pushed ever-higher taxes on corporations. One important significance of the taconite amendment, then, is not that it reveals an age-old hostility to taxes but rather that it illustrates the complicated "shifting of the burden" of taxation away from corporations and onto individual taxpayers in the second half of the twentieth century. By the end of the century, the modern conservative movement made hostility to taxes a central component of its political rhetoric, but in many areas such as the Iron Range, citizens were primed to receive the antitax message by a long history of tax policy that made them increasingly responsible for shouldering the fiscal burden of modern government.

The political transformations on the Iron Range were far more complicated than a straightforward narrative of industrial workers moving from liberal to conservative. Although many—perhaps most—miners supported the 1964 taconite amendment, a plan first proposed by conservative politicians, Iron Range voters did not permanently abandon liberalism or the DFL after the mid-1960s.

Indeed, throughout the last decades of the twentieth century, the Iron Range remained one of the most reliably Democratic areas in the nation. Behind the voting results, however, was a complicated reality. Industrial miners on the Iron Range may have followed their unions to vote Democratic, but they rarely followed the Democratic Party as it took on cultural and social issues at the end of the century. Iron Range journalist Aaron Brown notes the contradictory politics of the post-1960s Iron Range, and describes average Iron Range voters as "socially conservative, economically liberal populists who are averse to change."[82] The Iron Range sent politicians to Washington and St. Paul who were economically liberal but socially conservative. Much of this support was owing to the success of the taconite industry in maintaining a viable mining industry in the region. Iron Range voters remained liberal, in other words, because the mining economy kept going. Their quick abandonment of anticorporate radicalism during the taconite amendment debate, therefore, demonstrates the fragility of the labor–liberal alliance in the postwar era. Politically, labor liberalism was constructed on a foundation of ever-expanding industry. When deindustrialization and global competition kicked out those foundations, the entire edifice began to crumble.[83]

The taconite amendment revealed two overlapping political identities competing for the hearts and minds of industrial workers on the Iron Range in the postwar decades. The first was the identity of the liberal—sometimes even radical—working class. A dense nexus connecting liberalism, unions, and industrial work, this political identity had been forged in the heated labor battles of the early twentieth century and finished in the national political culture of the New Deal, which lionized industrial workers as central to the modern United States. What became clear in the postwar era, however, was that this political construction relied on a foundation of industrial growth for its survival. In the face of growing global competition and sweeping deindustrialization, a liberal working-class identity had few answers. Competing for the allegiance of the Iron Range's industrial workers was the identity of the conservative taxpayer. This identity emphasized industrial workers' role as citizens both paying for and consuming the services of government and, from that position, demanding accountability and value for their tax dollar. These identities were in constant flux on the postwar Iron Range and it is impossible to precisely locate where one ended and the other began. The taconite amendment was a small but significant window into the processes through which industrial decline and the threat of capital flight allowed the conservative taxpayer identity to make pragmatic economic arguments for industrial workers.

THREE

Taconite Bites Back

*The Reserve Mining Pollution Trial and
the Environmental Challenge to Industry*

When the Reserve Mining Company's E. W. Davis Works opened in
1955, the giant plant represented the triumph of technology as a
cure for industrial decline and overcoming natural limits. Taconite
ore was mined from the Peter Mitchell Mine near Babbitt, Minnesota, and sent by rail in piano-size chunks to the plant located in
Silver Bay on the shore of Lake Superior. There it was crushed and
ground to the consistency of flour, magnets separated the high-iron
magnetite from the surrounding siliceous rock, and the iron pellets
were baked hard and shipped down the Great Lakes to blast furnaces in Gary, Cleveland, and Youngstown. Creating the iron-rich
pellets—millions of tons of them each year—was the plant's reason
for existence. Yet something had to be done with the remainder of
the siliceous rock that did not contain iron and was useless to the
steel industry. The problem was an afterthought, but Reserve had
nonetheless come up with a solution that appeared ingenious in

85

1955: the waste rock—known as tailings—would be dumped into Lake Superior where currents would pull it to the bottom of the lake and forever remove it from sight.

The decision to dump tailings into Lake Superior, which seemed inconsequential in 1955, ultimately proved to be the plant's undoing. Beginning in the late 1960s and stretching through the 1980s, Reserve Mining Company's plant at Silver Bay was the focus of a drawn-out and complicated pollution dispute. Controversy over Reserve's tailings eventually led to the plant's shutdown by a federal judge, only to have other federal judges reopen the plant. More important, the case was a crucial development in the Iron Range's—and industrial America's—struggle to keep industry alive in the face of mounting pressures during the twentieth century, including the new environmental movement.

The Reserve Mining trial was extensively covered by journalists, environmental writers, and legal scholars. Existing accounts of the controversy emphasize the trial as a crucial development in the emerging environmental movement and environmental law. For legal analysts, the trial was significant because it provided the courts with crucial precedents for weighing the economic benefits of industry against harm to the environment and human health. As this chapter will demonstrate, the circumstances of the Reserve Mining trial put these contrasting values into sharp relief. As a journalist noted in 1974, "the federal court suit against Reserve Mining Co. has become the classic pollution case because it poses so sharply the questions of whether or not damage to a region's environment is worse than damage to the . . . economy."[1] Precedents set by the case proved important in later landmark environmental litigation, such as banning leaded gasoline in the United States. The trial is significant from a legal perspective alone.

Environmental historians, too, consider the Reserve Mining pollution controversy to be one of the most significant confrontations over industrial pollution in the late twentieth century. Environmental historians and environmental activists who have written about the case have emphasized how grassroots activists challenged a major employer and the political power structure to protect Lake Superior. The trial occurred just as the modern environmental movement was coming together in the United States and it became an important rallying point for many lawyers and activists in the new movement. From this perspective, standing up to Reserve Mining and forcing the company to stop polluting

Lake Superior was a foundational victory of the environmental movement.

Yet none of the existing accounts of the controversy have adequately understood its role within a longer history of the Iron Range and taconite's prior history as a technological fix for the region's declining iron-ore mining industry. Interpretations emphasizing the trial's importance to the legal system and the burgeoning environmental movement are certainly useful, but they suggest that it was only outside developments that challenged the older pattern of American industry. New developments occurring outside the Iron Range certainly did oppose older models of industrial growth, especially the environmental movement's rejection of the growth-at-any-cost ethos that characterized economic development in the mid-twentieth century. But there were also important yet overlooked ways in which the environmental challenges to industry, for Reserve Mining and elsewhere, arose from within industry itself. The question, then, is how did the prior history of using technology to promote economic growth and fight decline contribute to the new environmental challenges facing industry in the last third of the twentieth century?

From the perspective of the Iron Range's long fight against decline, the Reserve Mining pollution controversy reveals two important new challenges to industry that appeared in the late twentieth century. The first was the environmental movement itself, led by energetic, well-connected activists who increasingly challenged industry's right to freely pollute the air and water. Yet the organized environmental movement was only the tip of the iceberg for a wholesale shift in the value system of most Americans, who turned away from previous beliefs that development and growth should be privileged above all else. Slowly, many people came to believe that the harmful environmental consequences of pollution must be balanced against the economic good of mining and heavy industry. For many, the environmental costs of certain industries such as mining and smelting were simply too high to tolerate. Thus, those seeking to keep the Iron Range's mining economy alive in the late twentieth century now had to respond to a powerful new challenge: could new development be done in an environmentally responsible way? This challenge rippled through American industry in the last decades of the century.

The second challenge to industry arose from within. Especially for Iron Range miners and those working in heavy industry, environmentalism seemed to be an external force, rising up from an affluent urban middle class concerned with tourism and consumption. Indeed, there were real cultural divides between environmentalists and their opponents. Yet the Reserve Mining controversy suggests that environmentalism arose, in part, from developments within industry. Questions about the enormous tailings waste dumped into Lake Superior each day were not solely brought about by changing perceptions toward the environment. They were also brought about by the tailings themselves, perhaps inevitably making their presence felt in new ways after being displaced by the technological fix of taconite. Taconite, in other words, bit back. This suggests that the seemingly painless technological fixes used to prop up the mining industry earlier in the century were not, in fact, painless. They merely displaced the costs onto the environment in ways that would not be noticed for several decades.[2]

Background to the Reserve Mining Trial

Neither the use of Lake Superior's water in Reserve Mining's taconite milling operation nor dumping tailings into the lake were accidental. The plant had been designed to utilize Lake Superior as a source of freshwater and a sink for tailings waste. Reserve Mining Company's E. W. Davis Works, on the shore of Lake Superior in Silver Bay, Minnesota, used a tremendous amount of water in its taconite milling operation. The milling process designed years earlier by Edward W. Davis at the University of Minnesota's Mines Experiment Station depended on water throughout. Each ton of concentrated iron ore required approximately forty-five tons of water. This amount of water usage was typical for other taconite plants in northeastern Minnesota. Where the Reserve Mining plant differed from other plants, however, was its use of Lake Superior as a sink for dumping the plant's enormous volume of tailings waste. Taconite was a low-grade ore, meaning that each ton of rock mined from the ground contained approximately 22 percent iron. Almost two-thirds of the rock ended up as finely ground powder that was useless to the company and thus was dumped into the lake. The volume of tailings waste was staggering. During the plant's main years of operation from 1955 to 1980, the E. W. Davis Works dumped an average of sixty-seven thousand tons of tailings into Lake Superior each day. At the time of the plant's construction, Davis noted—with some satisfaction—that the plant would produce "enough [tailings] to cover 40 acres nearly 200 feet deep each year." As a comparison, historian Thomas Huffman notes that this amount was "more than two times the estimated solid waste garbage produced by New York City during the same period."[3]

The decision to use Lake Superior as a sink for tailings waste was a choice made early in the plant's design. When the Mesabi Iron Company built the first commercial taconite plant in 1919, the question of where to dump the tailings quickly emerged as a concern. Ultimately, the wet tailings were dumped in a swamp near the mill's isolated location in Babbitt. While experimenting with taconite at the Mines Experiment Station in Minneapolis, Davis and his staff had dumped tailings waste directly into the nearby Mississippi River. Yet these small operations were not comparable to the millions of tons of tailings later generated by the Reserve Mining plant. When the Reserve plant was being designed, engineers considered depositing the tailings on land near the mine at Babbitt. They quickly ruled out on-land disposal, however, because there was a limited water supply near the mine and, more important, it proved cheaper to haul the ore via rail to Silver Bay and dump the tailings in the lake.[4]

Harnessing hydraulic power to assist with mining had a venerable heritage prior to the mid-twentieth century. Many of the very first industrial technologies, such as the waterfall or the water-wheel, relied on hydraulic power to assist human production.[5] Hydraulic power had long been used both to separate minerals from tailings—recall the gold miner's simple water pan—and as a sink for tailings. Thus, Reserve Mining's use of hydraulic energy to dispose of tailings built on a very old tradition in industrial mining. Mines and mining had also long been associated with water pollution. Pumping water out of underground mines was one of the major challenges for early industrial mines. For most of mining's history, water pumped from mines or used to wash ore was dumped onto the ground or into a nearby stream. Water pollution was also among the earliest ways that the mining industry acknowledged the environmental consequences of mining. By the late nineteenth century, industrial mines were recycling their water supplies and carefully planning for water usage. Water conservation was motivated partly by economic considerations and fears of shortage in the dry mining regions of western North America, but there was an awareness that water pollution was a problem that accompanied mining.[6]

Davis and other Reserve Mining engineers chose to dump tailings into Lake Superior in the misguided belief that this was the most environmentally responsible means to get rid of tailings waste. They fell victim to the fallacy, common before the middle of the twentieth century, that tailings out of sight could not further harm human life or natural ecosystems. Describing the decision, Davis wrote: "we . . . concluded that the best place for fine taconite tailings was in the deep valleys at the bottom of Lake Superior. There they would be out of sight forever and posterity would not have to cope with them. We assured Reserve that the gray, sandy tailings of magnetic taconite would not in any way pollute the lake, interfere with any domestic water supply or with navigation, and would not adversely affect the fishing industry." This decision resulted from a series of experiments conducted at the Mines Experiment Station in which the staff built a scale model of Lake Superior and deposited taconite tailings into the model in the same proportion as the Reserve plant would deposit tailings into the lake. Later student researchers found that Davis's experiments contained a key flaw: tailings were suspended in cold water while the model held room-temperature water. The actual conditions in Lake Superior were reversed, with warm tailings entering the icy water of the lake. Although Davis would later be heavily criticized for his role in designing the system that dumped tailings into the lake, he designed this system in the misguided belief that the tailings would disappear beneath the lake's surface and would no longer trouble humans in the region.[7]

The Reserve Mining plant needed numerous permits to dump tailings into the lake and to use lake water in the milling process. The permits were issued with little contest in the late 1940s when the plant was under construction. Altogether, the Silver Bay plant required twelve permits from the state of Minnesota and the federal government. The permits fell into three main categories: permits from Minnesota's Department of Conservation allowing Reserve Mining to take water from Lake Superior for use in the plant, permits from the Minnesota Pollution Control Agency allowing the company to discharge water and tailings back into Lake Superior, and permits from the U.S. Army Corps of Engineers granting the company permission to construct a dock and breakwater for shipping. Public hearings were held to consider the proposed plant in 1947 and Reserve received widespread support based on the plant's promise of jobs and economic growth. Recalling the 1947 permit hearings, Edward W. Davis noted that the head of the Minnesota Department of Conservation "was very anxious to get something of that kind [a taconite plant] started." Overall, he recalled, "everybody was very helpful and enthusiastic about getting our plant started . . . Because it would employ three or four thousand men," a compelling argument in a region facing population loss owing to a lack of jobs. There was only minor opposition to the proposal from scattered outdoor enthusiasts worried about how tailings would affect Lake Superior fishing. This opposition was offset by the numerous conservation groups that supported the proposal, including the U.S. Fish and Wildlife Service and the Izaak Walton League. The necessary permits were granted in December 1947, officially allowing Reserve Mining to use 130,000 gallons of Lake Superior water per minute.[8]

When Reserve expanded the plant in 1956, the company requested amendments to the original permits allowing it to use more water than originally planned. Reserve was granted the amendment that gave it permission to use 260,000 gallons of water per minute. As was the case in 1947, there was little opposition to this increase in Reserve's use of Lake Superior water. When Reserve expanded the plant a third time in 1960, it again requested an amendment to use more water and discharge more tailings into the lake. The Minnesota Water Pollution Control Commission and the Minnesota Department of Conservation quickly granted Reserve the amended permits allowing it to use approximately half a million gallons of water per minute from the lake.[9]

From the perspective of 1960, Reserve had followed the letter and spirit of the law. The company had gained all the required permits for its taconite mill and it enjoyed widespread public support, both for the economic benefits it brought to the region and for its attention to natural resources. Reserve was proud of its record as a steward of the region's environment. In 1952, for example, it redesigned the harbor at Silver Bay to preserve a rare flower, the closed bluebell. The company, along with citizens' groups, pointed to the lake disposal of tailings as an environmentally responsible choice that removed the tailings from human view in perpetuity. Failure to see the danger of dumping sixty-seven thousand tons of tailings into the lake each day was only an extreme example of the era's wider failure to account for the natural world in its decisions regarding economic growth. The parties involved in the original decision—from Edward W. Davis to state pollution-control officials to citizens' groups—did not weigh the economic benefits of the plant against the environmental harm it would cause. Rather, they were blind to the possibility that it could cause environmental harm.[10]

Yet attitudes toward the environment were changing quickly. The years between the opening of Reserve Mining's Silver Bay plant in 1955 and the first alarms over pollution in 1968 witnessed a tectonic shift in how Americans thought of the natural world and their relation to it. As environmental historians such as Samuel Hays, Carolyn Merchant, J. R. McNeill, and others have shown, prevailing attitudes toward the environment shifted drastically in the post–World War II decades, moving away from a developmental ethos that emphasized economic growth and imagined the natural world—when it was considered at all—as a storehouse of resources to be managed by technology. In its place, many Americans began to think of the environment as a self-regulating system, autonomous from human control, that needed to be carefully preserved and shielded from human interference. Many Americans now thought of a pristine, healthy environment as a fundamental source of their quality of life. The words used to describe humans' relationship to the natural world reflected these changing attitudes: talk of "conservation" was replaced with "environment."[11]

Laws to limit water pollution were central to the environmental movement that developed in the 1960s and 1970s. For centuries, rivers, lakes, and oceans had been used by industry and urban populations as sinks for dumping waste. Yet the radically increased scale of industrial production in the twentieth century quickly overwhelmed the ecosystems of many bodies of water, leading to widespread pollution. Highly publicized incidents, such as when the Cuyahoga River in Cleveland was so polluted with industrial chemicals that it caught fire, drew the nation's attention to the reality that many waterways were no longer safe for drinking, swimming, or fishing. Public opinion shifted quickly, so that by the early 1970s, as historian Samuel Hays notes, "the prevailing sentiment now was that waste disposal was not a legitimate use of a stream." Water pollution was also among the most visible and easily understood types of environmental degradation. For example, educators throughout the nation encouraged schoolchildren to test water samples for turbidity, pH, temperature, and bacteria levels, quickly raising awareness of poor water quality among the nation's youth.[12]

At the national level, Congress passed keystone water pollution laws in the 1950s, 1960s, and 1970s. The result was that in two decades, beginning in the 1950s, American water pollution law was completely reshaped. What previously was understood as a strictly local issue was shifted to the federal level. The original basis for enforcing water pollution standards at the federal level was the Federal Water Pollution Control Act of 1948. The law was amended in 1952 and again in 1956, adding provisions that allowed the surgeon general to call enforcement conferences on specific pollution issues and mandated a timeline for polluters to clean up their wastes. Federal water pollution regulations increased again in the 1960s with passage of the Water Quality Act in 1965, which strengthened federal oversight of water pollution through the Federal Water Pollution Control Administration. Legislation in the 1970s further strengthened the federal role in ensuring that the nation's waters were clean and healthy. The 1972 Clean Water Act, the 1974 Safe Drinking Water Act, and the Clean Water Act of 1977 together gave the federal government, operating through the Environmental Protection Agency (EPA), the power to regulate drinking water and guarantee that American waters were safe for fishing, swimming, and drinking. For industry, these laws meant much stricter standards for treating wastewater and a new threat of lawsuits from environmental advocacy organizations, such as the Natural Resources Defense Council, that turned to the courts to enforce the new water pollution laws.[13]

Yet the federal water pollution legislation of this era was spurred more by political calculation than by grassroots pressure to clean up pollution. Historian Paul Milazzo has shown how Congress moved decisively to set the agenda on water pollution well ahead of public sentiment. More than the courts or any presidential administration, it was legislators, working through powerful standing committees, who were crucial to shaping the nation's water pollution regulations. And of all the congressmen who seized on water pollution as a political issue, none was more influential in the early years than John Blatnik, the Democratic representative for Minnesota's Eighth District, which encompassed much of northeastern Minnesota, including the Iron Range and Duluth.[14]

John Blatnik and the Origins of Water Pollution Regulation

If one man could encapsulate all of the tensions between industry and environmentalism in the postwar decades, it would be John Blatnik. The product of a typically poor, immigrant Iron Range

Congressman John Blatnik (center) *meets with iron-ore miners in 1954. Blatnik was deeply committed to the Iron Range's mining economy, a policy that ran into conflict with his water quality agenda during the Reserve Mining pollution lawsuit. Minnesota Historical Society Collection, por 4812 p1.*

mining family, Blatnik rose through Minnesota politics to become the U.S. representative for Minnesota's Eighth Congressional District. After his election in 1946, he served in Congress for decades, eventually rising to the top ranks of the Democratic caucus in the House of Representatives and serving as chair of the powerful Public Works Committee.

Blatnik dedicated his political career to two policies: economic development of his poor, rural district in northeastern Minnesota (a position that naturally led him to support taconite) and fighting water pollution. As he described in a speech prior to his retirement, "In my thirty-five years of public service I have been deeply concerned with two things—environmental conservation and economic development."[15] For much of his career, these legislative priorities did not intersect. But the Reserve Mining pollution trial brought them together with the force of a Shakespearean tragedy. The premier symbol of government-led economic development in the mining district, Reserve Mining's E. W. Davis Works, was directly polluting Lake Superior, perhaps the nation's most pristine body of freshwater. Blatnik's supporters saw the conflict between his two legislative priorities as a sad irony. His detractors saw evidence of something darker: a conspiracy between big business and politicians to pollute the lake and hide behind money and the law instead of facing the consequences of their actions.

Neither of these extreme views accurately captures the dynamic that led Blatnik to support both Reserve Mining and water pollution laws. For Blatnik, these two legislative priorities were different routes to a common goal of economic development spurred by a liberal state committed to redistributing wealth and conserving valuable natural resources. Blatnik's ideas about the appropriate role for government were developed as a young man growing up on the Iron Range during the economic ravages of the Great Depression. He developed a philosophy that government could, and should, take an active role in redistributing the nation's wealth from rich areas to poor communities. Like other New Dealers, especially those concerned with rural America, he believed there was a close connection between resource conservation and economic development. From this perspective, there was little contradiction between his support for taconite and his advocacy for tough water pollution laws.[16]

Blatnik's career illustrates an important, but often overlooked, continuity between the economic development policies of the mid-twentieth century and the environmentalism that blossomed in later decades. It was not coincidental that Blatnik promoted taconite plants and tightened water pollution laws. They both sought to use the power and money of the federal government to conserve resources while promoting the nation's economy. If there was a tragic element to Blatnik's career, it was that he could not foresee how these twin policies would ultimately work at cross purposes. The implication of this history, which is not limited to Blatnik and the Iron Range, is that the early environmental policies of the 1950s and 1960s often emerged from liberal politicians hoping to spur economic growth through the framework of the New Deal state's ability to redistribute money from federal to local sources. More important, they suggest how this approach to federally led economic development contained the seeds of its own collapse, simultaneously promoting industrial development that damaged natural resources and the tools for alerting the nation to the scope of that environmental degradation.

John Blatnik was born to Slovenian immigrant parents in the Iron Range town of Chisholm in 1911. He came of age just as the Great Depression hammered the Iron Range's mining economy, shutting down the mines and throwing thousands out of work. After finishing school, he took a degree from Winona State Teachers College. Like many young men from the mining district, he got a taste of New Deal liberalism while working for the Civilian Conservation Corps. He then started teaching in the Iron Range school system, quickly advancing through the school bureaucracy to become the assistant superintendent of the St. Louis County school system. Blatnik served in the Army Air Corps during World War II, where he was recruited into the Office of Strategic Services, a predecessor of the Central Intelligence Agency, and sent behind enemy lines in Italy and Yugoslavia.[17]

Returning to the Iron Range after the war, Blatnik immediately entered politics as a Democratic candidate for Congress. He had first joined the political arena immediately before the war when he was elected to the state senate. Early in his political career, he espoused the fiery, anticorporate rhetoric common to Iron Range radicals during the 1930s. In a radio speech, for example, he described the Iron Range as "a kingdom of steel autocrats who controlled production and prices and ruled labor with an iron fist . . . and as a result, the greatest natural resource, iron ore, the life bread of Minnesota, was stifled and crushed." His rhetoric had moderated by the time he entered Congress in 1946. He was one of many young Democratic war veterans elected that year, a cohort that included his friend John F. Kennedy. Blatnik, like Kennedy, was committed to continuing the New Deal's legacy amid the changing international scene of the postwar world.[18]

Blatnik's main priority when he entered Congress was economic development in his district. Although he supported numerous economic development measures, he was an especially strong supporter of taconite and Reserve Mining. He had worked with Edward W. Davis before the war to promote taconite and he was a strong backer

TACONITE BITES BACK

of the changed taconite tax law of 1941. By the late 1940s, as Reserve Mining was coming to life in Babbitt and Silver Bay, Blatnik became a close and powerful ally for the company. He supported Reserve Mining because he believed it would bring much-needed jobs to the Iron Range and, more important, he saw low-grade ore production as a way to preserve the mining economy into the future. He developed close ties to Reserve's leadership that certainly influenced his thinking. He was a personal friend of Reserve Mining's president, Edward Furness, and he befriended many other executives at the company, frequently spending vacations as their guest in northeastern Minnesota. These friendships also led to campaign contributions. Later analysis would find that "every operating officer of Reserve supported Blatnik with political contributions." Blatnik's close connection to Reserve Mining was not surprising given the intense political lobbying required to promote taconite (see chapters 1 and 2), but his ties to the company would haunt him when charges of pollution surfaced in later years.[19]

In Congress, Blatnik became famous for his leadership on behalf of water pollution laws. He was the author and key congressional supporter for several of the pioneering water pollution laws passed in the 1950s and 1960s, notably the 1956 amendments to the Water Pollution Control Act. These amendments, which allowed the surgeon general to call enforcement conferences in cases of pollution, put real teeth into federal water pollution laws for the first time. His leadership on clean water legislation was widely recognized among fellow politicians and early environmental groups. For instance, the Izaak Walton League added Blatnik to their honor roll in 1960, citing his "outstanding service in the cause of conservation" and his "relentless work on behalf of clean water for all Americans." In 1964, Vice President Hubert Humphrey described Blatnik as "the daddy of all these Water Pollution Control measures. Long before anyone else even lifted a finger to do something about this problem of pure water and keeping our streams and lakes clean, John Blatnik moved in and took the leadership."[20]

On the surface, it would appear that Blatnik's support for clean water laws was antithetical to his emphasis on industrial and economic development. Yet closer analysis of his ideas and motivations suggests that he turned to clean water as a new means to support economic development and dole out federal money to localities. As historian Paul Milazzo argues, Blatnik, like many of the Democratic congressmen who wrote the environmental legislation of the 1950s and 1960s, was an "unlikely environmentalist." Prior to the mid-1950s, there is little evidence that Blatnik was concerned, or even aware, of water pollution as a legislative matter. As Milazzo writes, "Blatnik had about as much interest in championing water pollution control as most of his colleagues, which is to say, not much." Blatnik was noticeably disinterested in water pollution bills when his Rivers and Harbors Subcommittee of the House Committee on Public Works first took up the matter in 1955. He only took an interest in water pollution after a Republican congresswoman unexpectedly sent a modest water pollution bill down to defeat in his subcommittee. Blatnik was personally slighted by the defeat and the corporate interests that he believed had killed his routine bill. Yet once his interest was piqued, he realized that water pollution legislation offered an excellent vehicle to expand his mission of economic development and the distribution of federal funds.[21]

The water pollution laws spearheaded by Blatnik were impressive legislative vehicles for shifting federal funds to localities throughout the nation. In essence, they were legislative pork. By doling out federal money for water treatment facilities, Blatnik's water pollution laws of the 1950s revived the federally funded but state-led infrastructure building that had characterized the 1930s. His programs were, not surprisingly, wildly popular among small-town mayors and organized labor because they paid for infrastructure and construction jobs that would have been out of reach for many small towns without federal support. In particular, many small towns would not have been able to afford expensive water treatment facilities without extensive financial support from the federal government. Groups such as the American Municipal Association and the U.S. Conference of Mayors were enthusiastic supporters of Blatnik's bills. In return for their support, Blatnik altered his bills to ensure that large metropolitan areas would not take all of the available grant money. The result was that many small towns and cities took up antipollution measures not because of genuine worries about pollution, but because they wanted a piece of the federal money. As Blatnik's aide noted, the congressman's subtle maneuvering "made pollution controllers out of every member of Congress who had a channel to dig, a harbor to deepen, a bridge to build, [or] a post office to name."[22]

Although legislators around the country tapped into the pollution-control money for their own pet projects, Blatnik was especially pleased to bring some of the federal funds back to his Minnesota district in the form of a cutting-edge scientific laboratory for water testing located in Duluth. The Duluth National Water Quality Laboratory was authorized by the Federal Water Pollution Control Act Amendments of 1961 and the 1963 appropriations act. Although the lab's location in Duluth ostensibly would allow it to monitor Lake Superior's water, many understood that the lab was a pork-barrel project designed to bring federal money and jobs to Blatnik's Eighth District. Duluth residents, many of whom were grateful for the 140 well-paid jobs the lab brought to the city, referred to the facility as "Blatnik's lab." Ironically, "Blatnik's lab" would come to initiate the water pollution research that would eventually bring down his legislative career.[23]

Blatnik's approach to water pollution was challenged in the 1960s by an environmental movement that was less willing to work within the framework of the New Deal legislative system. Increasingly, environmental groups challenged the cozy relationship between conservation and industrial development that existed in Congress. By the mid-1960s, Blatnik had developed a national reputation for his leadership on water pollution issues and he was eager to maintain that reputation and the prestige associated with it. Yet the emerging environmental movement of the 1960s was far less enamored with him and his legislative approach to solving pollution problems. Many leaders of the environmental movement considered Blatnik and his Public Works Committee to be conservative in pushing for regulations on polluters. Blatnik favored local control of pollution issues over national regulations, which gave him powerful political leverage. But environmentalists argued that turning pollution regulation over to local governments allowed powerful industries to pollute with impunity because cities and towns were often hesitant to regulate industries that were major employers. This was a reasonable criticism of Blatnik's priorities, but as the Reserve Mining case developed, some environmentalists involved with the trial became convinced that Blatnik himself was behind the pollution. For example, Grant Merritt, a leader of the environmentalists opposing Reserve Mining, singled him out as a powerful protector of Reserve Mining. Merritt proposed that Reserve's tailings delta in Lake Superior be named Blatnik State Park. He also accused Blatnik of being drunk during important meetings and physically threatening environmental activists. "He was used to throwing people around," Merritt said, "you know, like a lot of those Iron Range types in the legislative process." Merritt's comment hints at the growing animosity between environmental activists and the Iron Range residents who sided with industrial development over antipollution regulation. Blatnik reciprocated in his dislike for Merritt, once calling him "sort of a neurotic" who "didn't need scientific evidence" to prove that pollution was happening.[24]

Stoddard Report and Enforcement Conferences

After years of muted concern among North Shore residents and state and local officials, Reserve Mining's dumping of tailings into Lake Superior exploded into a national controversy in the late 1960s when the press obtained a Department of the Interior report claiming that Reserve Mining was polluting the lake. The report, and the ensuing controversy over its leak to the press, drew national and international attention to Reserve's discharge. The report was the first step in what would become a decade of litigation and regulation.

Regulation for Reserve's discharge was primarily a state issue prior to 1967. The majority of Reserve's dumping permits were issued by Minnesota state agencies. The exception was a minor permit issued by the U.S. Army Corps of Engineers that gave Reserve permission to build a dock and breakwater at Silver Bay and deposit tailings at the site. The permit was necessary under an obscure section of the 1899 Rivers and Harbors Act that allowed the Corps to regulate polluters, but it was rarely enforced prior to the 1960s. By the late 1960s, though, attitudes toward water pollution were changing within the federal bureaucracy. As early as 1963, Wisconsin senator Gaylord Nelson expressed concern over Reserve's discharge and urged the federal government to take action because the pollution crossed state and international boundaries. Given increasing concern with pollution, it was not surprising that Reserve's dumping attracted the attention of federal regulators in the late 1960s.[25]

The year 1967 proved to be pivotal in the Reserve pollution saga. That year, Reserve applied for an extension of the federal permit allowing the company to discharge tailings waste into Lake Superior. Although previous extensions had been routine, several offices within the Department of the Interior now objected to an extension. As a result, the department launched a study group charged with investigating how taconite tailings were affecting Lake Superior. The study group was headed by Charles Stoddard, who had taken charge of the Duluth National Water Quality Laboratory several years earlier. Stoddard proved to be a fateful choice. He was a longtime environmentalist and had influential connections within the Department of the Interior, especially with Secretary Stewart Udall. In 1968, Stoddard brought together officials within the Department of the Interior, the Minnesota Pollution Control Agency, the Minnesota Department of Conservation, and the Wisconsin Department of Natural Resources under the umbrella of the Taconite Study Group. The Taconite Study Group's report—commonly known as the Stoddard Report—landed like a bombshell in the Department of the Interior and among Lake Superior area residents. The report claimed that Reserve Mining's tailings were definitely polluting Lake Superior and affecting fish and wildlife in the lake. Further, the report said it would be technically feasible and relatively inexpensive for Reserve Mining to deposit its tailings on land rather than in the lake. For the first time, the federal government was on the record in claiming that Reserve was polluting the Great Lakes by dumping taconite tailings.[26]

Stoddard's report provoked intense controversy within the federal government. Although environmentalists within the Department of the Interior were eager to stop pollution, Reserve Mining had influential allies throughout the legislative and executive branches who were deeply worried about the federal government moving to stop the company's production at Silver Bay. Representative Blatnik was very upset when he learned about the report's findings. Although the exact sequence of events is unclear, Blatnik told Stoddard that more study was needed and urged him not to release the report to the public. There were allegations that he went further, complaining to his liaisons in the Department of the Interior about Stoddard and the report. Political scientist Robert Bartlett, who conducted the most authoritative review of Blatnik's actions in response to the report, argues that the precise sequence of events is "not easily reconstructed." Blatnik and his staff clearly urged more study of the issue before releasing any official reports and, ultimately, the report was hidden within the Department of the Interior. The Stoddard Report was never accepted as an official report by the Department of the Interior, despite the fact that it was the impetus for all subsequent federal action. Bartlett concludes that Blatnik did not directly threaten or coerce Department of the Interior officials. He did not "pound the table or make blatantly improper requests." But it is undeniable that Blatnik's opposition to the report shaped the actions of government regulators. Blatnik led a powerful congressional committee and he clearly did not want widespread controversy surrounding Reserve Mining. He was not the only person critical of the report. Officials in Minnesota who had studied Reserve Mining and its discharge for years also argued that the report went too far in alleging that Reserve was undeniably polluting the lake. Complicating matters was the fact that the report was finished just as the incoming Nixon administration was reshuffling high-level Department of the Interior appointments. Stoddard, for example, was a political appointee and he resigned his post after the 1968 election.[27]

Worried that the incoming administration would shelve the report, Stoddard leaked it to a Minneapolis journalist. On January 16, 1969, the *Minneapolis Tribune* ran a front-page story on the report's findings. The general public in Minnesota was now alerted to the potential ecological harm being caused by taconite tailings in the lake. As one observer noted, "the calculated leak of the [Stoddard Report] triggered an avalanche of criticism of Reserve by state and federal regulatory agencies, conservation groups, and the public." Newspapers throughout the nation picked up the story of a single company that threatened the largest and cleanest of the Great Lakes, ultimately launching the issue into the court of public opinion.[28]

Following the controversy of the Stoddard Report, Reserve's tailings and their impact on Lake Superior were debated at a series of enforcement conferences that brought together government regulators, Reserve Mining Company, and concerned citizens to determine the scope of the problem and potential solutions.

The enforcement conferences proved to be an awkward forum that left few people happy with the outcome and allowed Reserve Mining to delay any meaningful action toward cleanup or limiting discharge. The reasoning behind the conferences was that they would be a neutral forum to consider all relevant scientific evidence before making a decision about pollution abatement. The conferences were a vestige of Blatnik's original 1956 water pollution legislation. When he first proposed the bill in Congress, he wanted harsh and decisive regulation following a government finding that a company was polluting. But the bill's enforcement provisions were drastically reduced as a result of stiff opposition from other representatives. The enforcement conferences that ultimately emerged from the bill were far less punitive and were intended to create consensus on the existence and extent of pollution prior to a lawsuit. Although spurred by good intentions, in practice the enforcement conferences often allowed industrial polluters to delay cleanup indefinitely. Milazzo argues that, "with all the . . . procedural delays, the chances of ever taking a polluter to court had actually grown more remote . . . Only one court action was ever filed under the act in the decade following its passage." By the 1970s—partly in response to the Reserve Mining example—lawmakers realized that the enforcement conferences merely led to drawn-out enforcement of the law.[29]

The first enforcement conference opened in a lavish ballroom at the Duluth Hotel on May 13, 1969. More than four hundred onlookers packed the room throughout the proceedings. Outside, environmental groups and high-school students carried placards protesting Reserve's dumping. Blatnik officially opened the conference. Torn between his reputation as the congressional father of water pollution legislation and his support for the taconite industry, he carefully noted that the federal government's jurisdiction was limited in this issue. Never missing an opportunity to highlight his accomplishments, he also touted the National Water Quality Laboratory's ability to find and solve any possible water pollution. Over three days, the conference listened to testimony from more than 120 witnesses. An enormous volume of testimony was accepted during the conference, including what one reviewer noted were "all sorts of documents of questionable relevance, importance, or validity . . . some of them quite lengthy." Despite three days of dense testimony, the first enforcement conference produced little concrete evidence that Reserve was polluting the lake, although extensive discussion of the enormous volume of Reserve's tailings led most observers to the commonsense conclusion that this huge amount of waste must somehow be harming Lake Superior's water quality. The first enforcement conference adjourned on May 15, 1969. No action was planned against Reserve.[30]

The conference was notable, however, for offering the first glimpses into Reserve Mining's defense strategy. In the face of mounting scientific evidence that tailings were harming Lake Superior and growing public outrage over the scale of the dumping, Reserve countered with its own scientific findings suggesting that tailings were not polluting the lake. Reserve's defense was coordinated by William K. Montague, a longtime attorney for the iron-ore mining industry in northeastern Minnesota, and his young partner, Edward T. Fride. During the conference, water scientists testifying on behalf of the company argued that the tailings were definitely not polluting the lake. Reserve Mining's president,

Edward M. Furness, also defended the company and its environmental record. Furness described the tailings as similar to inert sand that was little different from the bottom of the lake. He also highlighted the company's economic benefit to the region, noting that Reserve Mining paid millions of dollars of tax revenue to the state. Scientific uncertainty over whether tailings were polluting the lake and highlighting the company's economic benefits would become the main strategy of Reserve's defense in coming years.[31]

Between the end of the first enforcement conference and the beginning of the second, scientists at the National Water Quality Laboratory launched an intensive investigation of Lake Superior to gather more scientific data on tailings. Between May 1969 and August 1970, scientists produced numerous experiments that demonstrated that Reserve's tailings were creating large patches of green water and contributing to algae buildup that was prematurely aging Lake Superior. When the second enforcement conference began at the Duluth Hotel on April 29, 1970, the scientific and political tide had turned against Reserve Mining Company. Scientists working for the EPA were gathering more and more data showing that the tailings had a demonstrable effect on Lake Superior's water quality, and several Minnesota politicians took up the antipollution cause. Reserve continued to argue that its tailings waste was not affecting the lake, but the company's arguments were increasingly less convincing. After listening to Reserve's arguments, the chairman of the enforcement conference mockingly revised Churchill's famous quote: "Never has the discharge of so much done so little."[32]

By late 1970, Reserve Mining grudgingly accepted that it would need to change its disposal method away from simply dumping tailings into the lake at Silver Bay. The major issue at this stage was whether the company would shift to on-land disposal or a modified method of discharging tailings into the lake. Reserve clearly favored continued disposal into the lake. Rather than just pouring the tailings into the lake at the Silver Bay tailings delta, however, the company proposed to build a deep pipe that would carry the tailings far below the lake's surface. The so-called deep pipe plan also added thickeners to bind the finely ground tailings together into a slurry that would hopefully form a sandy reef offshore. Presenting the plan to the enforcement conference, Furness argued that the pipe would carry tailings to a deep trough in Lake Superior, where the tailings would settle and not disperse throughout the lake.[33]

Environmental groups, the EPA, and state regulators rejected the

A delta of tailings spreads out from the Reserve Mining Company's E. W. Davis Works into Lake Superior, circa 1962. Initially noncontroversial, Reserve Mining Company's dumping of tailings directly into Lake Superior spurred a long and contentious pollution lawsuit. Minnesota Historical Society Collection, HD3.27 p15.

deep pipe plan immediately. By 1971, Reserve's opponents wanted the company to stop dumping tailings into the lake altogether and pushed for on-land disposal. At the enforcement conference, EPA regulators argued that the deep pipe plan clearly would not reduce the overall volume of tailings waste flowing into the lake and the addition of chemical settling agents would cause "unknown ecological impact." The plan did nothing to eliminate the concerns over green water and suspended solids that spread throughout the lake as a result of the tailings. On-land disposal of tailings was the clear preference of government regulators. Every other taconite plant on the Iron Range used on-land disposal, they argued, so on-land disposal was technically feasible and economically viable. Reserve Mining, however, was dead set against on-land disposal. The company argued that on-land disposal of tailings would require an enormous, costly, and complicated system of dams to create a holding pond, as well as create blowing tailings dust that would spread across the North Shore. Dumping tailings on land versus in the lake would cost more than $200 million and add approximately $2.50 to the cost of each ton of ore that the plant produced. By the spring of 1971, then, both sides were dug in to their opposing positions. Government regulators wanted the company to stop dumping in the lake altogether and switch to on-land disposal, and Reserve Mining was equally insistent that it continue to dump into the lake. Faced with hardening positions, the chairman of the Federal Water Pollution Control Administration recommended that the government end the enforcement conference on April 23, 1971. The chairman proposed that the government file suit against Reserve to stop the tailings discharge. As stipulated by federal law, Reserve Mining Company then had 180 days to develop an alternative to lake disposal or face a lawsuit from the Department of Justice. The administrative portion of the controversy ended as the issue was taken up in federal court.[34]

The Federal Trial and Asbestos Controversy

Controversy over Reserve's dumping into Lake Superior now shifted from the clunky enforcement conferences to a federal district courtroom. The trial would prove to be one of the longest, costliest, most complex, and most controversial environmental battles of the twentieth century. When the actions of state courts and appeals are included, the lawsuit lasted for more than a decade and brought decisions from Minnesota district courts all the way to the U.S. Supreme Court. The rulings that emerged from the lawsuit also proved to be crucial precedents for later environmental law. Legal scholar Daniel A. Farber argues that the case "was the first major judicial confrontation" over assessment of environmental risk versus economic benefit. "It remains a leading case on the subject of risk regulation" in the American federal courts. Later pollution decisions, such as *Ethyl Corporation v EPA*, which justified regulating leaded gasoline because of potential health risks, were based largely on precedent from the Reserve Mining decision. Journalists observing the case agreed that it was enormously important. One *Chicago Tribune* reporter said that trial was no less than "the most important environmental suit ever tried."[35]

Lawsuits against Reserve began in 1969 when the Sierra Club sued the state of Minnesota in an attempt to force the state to enforce its own water quality laws. Reserve Mining immediately filed numerous countersuits and the case came before a state court. On December 15, 1970, a state judge ruled that Reserve was likely polluting Lake Superior, or at least contributing to the green water phenomenon. The judge ordered Reserve to begin considering alternative methods to dispose of its tailings. Thus, there were parallel cases developing in state and federal courts throughout the early 1970s. Bouncing back and forth between jurisdictions surely made the case lengthier and more complicated than it otherwise would have been.[36]

The EPA's formal complaint was filed on February 17, 1972. Following the breakdown of the enforcement conferences, the EPA asked the Department of Justice to file suit against Reserve Mining. The initial complaint named only Reserve Mining Company as a defendant, which was controversial because some environmental advocates urged the government to include Reserve's parent companies, Armco and Republic Steel, as defendants in the hope that this would bring pressure on the powerful steel industry. For instance, one EPA attorney was worried that the government's case against Reserve was "thin" and urged the government to launch an all-out legal assault by suing Reserve and both parent companies for violating every federal pollution statute possible. The government chose instead to limit its complaint. At this point, the government insisted only that Reserve "abate the pollution from the discharge of taconite ore waste into Lake Superior" and did not specifically insist on an on-land disposal solution. When pretrial hearings began, the federal government, represented by the Department of Justice, was joined by several environmental advocacy groups, including the Save Lake Superior Association and the Northern Environmental Council. Michigan and Wisconsin joined the suit as well, and Minnesota would join several months later after concluding the existing case in state court. Opposing the government was Reserve Mining, which was joined by several development agencies, including the Duluth Area Chamber of Commerce and the Northeastern Minnesota Development Association.[37]

Judge Miles Lord was selected to preside over the federal case. Lord had a long history with iron-ore mining and the Iron Range that would prove fateful in his handling of the case. He grew up near Crosby, Minnesota, on the Iron Range, before working his way through college and law school at the University of Minnesota. Lord's career began in politics rather than the judiciary. Running as a DFL candidate, he was elected as Minnesota's attorney general in 1954. Like most prominent Minnesota politicians of the 1950s and 1960s, he was wooed by Reserve Mining during the taconite amendment debate (described in chapter 2). During that time, Lord joined Hubert Humphrey for a vacation near Silver Bay with a Reserve Mining executive. The trip convinced him to support the taconite amendment. After his appointment as a federal judge, however, he developed a reputation as an unapologetic judicial activist who was willing to aggressively defend average citizens against corporate power. Throughout the trial, he attracted nationwide praise for his aggressive defense of the environment. Opponents of judicial activism—which included the Eighth Circuit Court of Appeals— saw Lord as an example of a judge who was more advocate than impartial mediator. As one legal scholar claimed, "In the view of many lawyers, Judge Lord was on the bench to do justice; he did not allow anything to stand in the way—whether it was Congress, appellate courts, or the evidence in a case."[38]

The trial, *United States of America v. Reserve Mining Company,* opened in the Minneapolis federal courthouse on August 1, 1973. Initially, the case hinged on whether Reserve Mining and its taconite tailings were in fact polluting Lake Superior. This proposition was more complicated than it seemed, for there were few concrete scientific data demonstrating that the tailings were a harmful pollutant and not just inert sand. The government's case instead relied on a commonsense notion that dumping so much crushed rock—tens of thousands of tons each day—into the lake had to be harmful. Describing the government's initial legal strategy against Reserve, one trial lawyer wrote: "Our position would be that Lake Superior was nearly perfect and that any change in the lake had to be bad." Legally, however, the government was required to provide scientific proof of pollution. To do this, the government brought in numerous expert witnesses who established that Reserve Mining's tailings were the only possible source of the crushed rock that was polluting Lake Superior.[39] Yet all of the aquatic pollution questions would soon be overshadowed by the threat of a public health crisis.

Before the federal trial began, the entire case was upended by revelations that taconite tailings in Lake Superior might be causing cancer. At issue was a potential link between taconite tailings and asbestos. The asbestos connection first came to light in late 1972. During a meeting in Duluth, a local geologist informed Arlene Lehto, one of the most active local environmentalists working against Reserve, about recent findings that asbestos-like fibers found in Japanese rice might be linked to stomach cancers. The geologist told Lehto that Reserve's tailings contained amphibole fibers—part of the family of minerals related to asbestos—that looked much like the fibers found in Japanese rice. Ominously, these fibers appeared in Duluth's water supply. After an evening of research, Lehto publicized the possible connection between taconite tailings and asbestos. Speaking to a union meeting, she argued that drinking water with taconite tailings might cause cancer. Revelations that Reserve's tailings were not only polluting Lake Superior but might be poisoning local water supplies fundamentally reshaped the case against the company.[40]

Scientists soon confirmed that there were ominous similarities between taconite tailings and asbestos fibers. Philip Cook, a researcher at the Duluth National Water Quality Lab, quickly followed up on Lehto's argument that Reserve's tailings were carcinogenic. He found that Reserve's tailings were laden with cummingtonite-grunerite and that this mineral, when finely crushed, appeared very similar to asbestos fibers. Delving into the scientific literature on asbestos, Cook realized that amphibole fibers from crushed cummingtonite-grunerite were linked to lung cancer when people were exposed to them in the air, but there were no existing studies of how the fibers might affect those who drank water contaminated with them. In June 1973, officials from the EPA, the Duluth National Water Quality Lab, the Department of Justice, and the Food and Drug Administration met in Duluth to discuss the asbestos issue. Despite the lack of scientific consensus over whether the fibers were harmful, officials concluded that "drinking the amphibole particles in the Duluth water would create a health risk of undefined proportions." More ominously, all the officials agreed that they would not drink the water if they lived in Duluth. Government regulators elected to take their concerns to the public in June 1973, despite conflicting evidence over the health risks posed by taconite tailings. On June 15, the public learned that Reserve's tailings contained asbestos-like fibers that might be causing cancer via the Duluth water supply. Reserve Mining Company was invited to sign the announcement, but declined. Although everyone hoped that the news would not cause undue alarm, newspapers throughout the Lake Superior region blasted the news across the front page in large headlines.[41]

The court responded immediately by ordering scientific studies to determine whether taconite tailings were contributing to cancer. Irving Selikoff, a doctor in New York City and one of the nation's foremost medical authorities on asbestos-related cancers, launched an immediate study of deaths in the Duluth area. If taconite tailings were carcinogenic, he reasoned, then the rates of lung and stomach cancers in Duluth and the North Shore should already be higher than average. The studies, conducted with great haste, proved inconclusive. Autopsies from recently deceased Duluth residents did not find high concentrations of amphibole asbestos fibers. More conclusive medical studies would take much longer, so it was impossible to rule out the cancer threat. Selikoff also agreed to act as a key witness for the prosecution, testifying that Reserve's tailings discharge was "a distinct public health hazard" and "a form of Russian roulette."[42]

The cancer scare over taconite tailings appeared just as Americans were learning of the health risks from asbestos. In 1970 and 1971, scientific evidence connecting asbestos to cancer first came to light as a widespread public health concern. Scientific studies of workers in an asbestos-products factory revealed much higher than expected rates of lung and stomach cancers. Public alarm spread following journalist Paul Brodeur's exposé of the asbestos industry, *Expendable Americans*. The book revealed enormous public health risks from asbestos and suggested that the industry had known about these risks for years but covered them up. Yet the precise risks were unknown at the time. For many Americans, dawning knowledge of asbestos in the early 1970s hinted at a toxic environment and powerful corporations that could not be trusted to do what was best for public health. Thus, the cancer scare and the entire Reserve Mining case occurred during a moment when Americans were first coming to grips with the reality that the modern American lifestyle produced hundreds of carcinogenic substances. Those involved in the Reserve case came to view the trial through this prism of a landscape filled with unknown, but potentially catastrophic, public health risks.[43]

Residents of the small North Shore communities such as Silver Bay either dismissed or ignored the warnings. Public health officials braced for widespread panic in the small towns closest to Silver Bay that drew water from Lake Superior, but residents were unfazed. In Silver Bay, some residents thought the announcement was a hoax by the EPA designed to scare away tourists. Others dismissed it as a temporary panic that would soon be discounted.

Silver Bay's mayor told a reporter, "we feel it [the drinking water] will be proven harmless, just like everything else that's come up." Others believed that the medical fears were overblown, joking that Dr. Selikoff should be named "Dr. Sillycough." In private, however, Silver Bay residents expressed more concern than in their statements to reporters. Few would speak candidly to reporters from outside the area. When a *New York Times* reporter arranged a secret meeting with one Reserve Mining worker outside the town, the worker admitted that he was, in fact, deeply worried about the health effects of working at the taconite plant.[44]

In Duluth, on the other hand, many reacted with alarm over the possibility that their tap water might contain carcinogenic fibers. Duluth, the largest city on Lake Superior, was especially vulnerable to water pollution in the lake. Lake Superior's water was of such high quality that the city had not installed a water treatment facility prior to the 1970s. Water pumped from the lake was only disinfected before it was used in the city's water supply. After the announcement of pollution, several thousand Duluth residents took matters into their own hands to secure safe drinking water. They traveled to friends and family who had wells. A local television station that had its own well allowed people to take water home. Some Duluth residents installed filters on their taps.[45] Those with young children were especially concerned that exposure to asbestos-like fibers might be harmful. One Duluth resident who drove eight miles to pick up safe drinking water said he was doing it for his young son: "I don't think it's worth the gamble on my son's medical future to drink Duluth tapwater." Yet, many other residents patiently waited for more facts before taking action. As one taxi driver put it, "let's wait and see what this doctor [Dr. Selikoff] comes up with." The federal trial taking place in Minneapolis thus occurred against this background of mounting worry in Duluth and the mining region.[46]

Revelations that Reserve Mining's tailings might be carcinogenic fundamentally altered the trial from a lawsuit over environmental degradation of Lake Superior to an urgent matter of public health. From this moment, the case revolved around the possibility that Reserve's tailings would cause mass sickness and death in northeastern Minnesota. Judge Lord described the finding as "potentially . . . the number one ecological disaster of our time."[47]

The federal government opened its case against Reserve by presenting a litany of expert witnesses who testified that the company was definitely polluting Lake Superior with crushed cummingtonite-grunerite and possibly poisoning anyone living near the lake with carcinogenic fibers. Expert witnesses were presented to prove both of these claims. Water and mineral specialists first established that Reserve Mining's tailings were the only possible source of the cummingtonite-grunerite found throughout the lake, countering Reserve's claim that the mineral might be entering the lake from natural sources such as a stream. Next, the government brought public health experts to the stand to argue that asbestos-like fibers found in the lake were a public health risk. Reserve Mining argued that its tailings were not polluting the lake in any way. The tailings it dumped into the lake were nothing more than inert sand that was little different from the lake bed, the company's lawyers argued. Interestingly, by insisting that the tailings were not pollution, the company prohibited itself from making the more complicated, but perhaps more compelling, case that pollution from mining was inevitable and must be balanced with its economic and social benefits.[48]

Early in the trial, Judge Lord and the federal government floated the idea for a possible settlement of the case if Reserve Mining would quit dumping in the lake and begin on-land disposal of tailings. These discussions never advanced far, however, because Reserve Mining was committed to the deep pipe plan to discharge tailings near the bottom of the lake. The company refused to even consider the possibility of on-land disposal. It argued that relocating its tailings to on-land disposal near the mine at Babbitt, as the prosecution wished, would be prohibitively expensive, and estimated the cost of such a move to be $574 million. Prosecutors thought that this amount was artificially high because several other taconite plants operated on-land disposal operations at far lower costs. Judge Lord, worried about the possibly carcinogenic effects of the tailings in the lake, refused to consider any disposal option that left tailings in the lake.[49]

Settlement talks finally advanced in the spring of 1974, when government lawyers discovered an internal technical document from Reserve Mining showing that the company already considered the deep pipe plan to be technically impossible. The document further noted that it would be possible for Reserve to relocate its crushing and grinding operations from Silver Bay to the mine at Babbitt. According to the report, crushing and grinding operations had been moved from Babbitt to Silver Bay for financial reasons, not technical problems. The engineer's report noted that other taconite plants operated near Babbitt using on-land disposal and even described specific areas that could be used for dumping tailings on land. More damning for Reserve were internal documents subpoenaed from Armco and Republic Steel showing that Reserve Mining had rejected the proposed deep pipe plan as technically unfeasible several years earlier. Judge Lord was now upset with Reserve Mining, believing that the company had wasted the court's time arguing over a proposal that it had already rejected. For journalists following the case, the revelation that Reserve Mining had already deemed the deep pipe plan to be unfeasible and was studying on-land disposal options, while claiming in court that on-land disposal was impossible, reframed a long, technical trial as a juicy "corporate cover-up."[50]

Recognizing that Judge Lord and public opinion were turning against the company, Reserve Mining's attorneys quickly pushed for renewed settlement talks. The company was now resigned to an on-land tailings disposal option. Arguments that on-land disposal was too expensive and too difficult had been thoroughly refuted by the government's expert witnesses and the damning internal documents. Changes in the iron-ore market also helped Reserve to accept an on-land disposal option. Just as Reserve and the federal government were debating possible locations for on-land tailings disposal, the price of iron-ore pellets shot up dramatically. The higher prices made Reserve more comfortable in accepting a more expensive disposal plan. Armco Steel and Reserve Mining chairman C. William Verity claimed that Reserve's goal was now to move toward on-land disposal "on the most favorable terms possible." Verity, who quickly became a key witness in the case, was an influential Republican insider and conservative activist. As the case developed, Verity and Judge Lord became increasingly hostile to each other.[51]

Settlement talks stalled after only a few meetings. The EPA wanted Reserve Mining to shift all production, including tailings disposal, to its Babbitt mine. Reserve proposed to dump tailings at an on-land site near Silver Bay and would not promise any reduction in air pollution. Frustrated with the stalemate, the EPA authorized its attorneys to press for an injunction to stop Reserve's pollution entirely.[52]

The long trial reached a climax on April 20, 1974. Judge Lord lectured Verity that he was tired of Reserve's stalling tactics and told the defense that they had the lunch hour to bring forward a proposal for settlement or risk having the plant shut down by judicial injunction. After lunch, Verity proposed that Reserve would begin engineering an on-land disposal site near Silver Bay, but only on the condition that it could continue operations until the new tailings impoundment was ready, that the federal government would provide any necessary financial assistance, and that there would be no further legal claims over Reserve's pollution. A courtroom observer called Verity's statement "the most amazing display of arrogance I've ever seen . . . it was like Armco and Republic were giving [Judge Lord] the finger right in his face." Later that afternoon, Judge Lord delivered his opinion: Reserve Mining was polluting Lake Superior with asbestos-like fibers, those fibers were a health hazard to people living near the lake, Reserve's discharge of fibers into the air was a similar health hazard, and, ultimately, he enjoined Reserve Mining to stop all discharge into the air or water as of 12:01 a.m. the next day. After years of studies, conferences, and trials, the E. W. Davis Works of Reserve Mining was silenced by judicial order on Sunday, April 21, 1974.[53]

Almost immediately after Judge Lord's shutdown injunction, Reserve Mining's attorneys appealed the decision to the U.S. Court of Appeals for the Eighth Circuit. At the time, the Eighth Circuit judges were attending a conference in Springfield, Missouri. They met hastily in a hotel room on April 22. After hearing only thirty minutes of argument, they quickly reached a verdict to stay Judge Lord's ruling and allow the plant to resume operations until the trial could be considered by the appellate court. In May and June 1974, the Eighth Circuit Appellate Court heard Reserve's appeal of Judge Lord's ruling in more detail. The appeals court ruled that all evidence of medical harm was unproven and granted another seventy-day stay of Judge Lord's shutdown ruling. But the court recognized that dumping tailings in the lake must stop and ordered Reserve to submit a new proposal to dispose of tailings on land and cut back on air pollution. Lawyers for the prosecution appealed the decision to the U.S. Supreme Court, but the Court declined to hear the case. By this time, the government's case was complicated by the fact that it reached the appeals court just as Watergate was reaching a crisis point in the Nixon administration. For the government prosecutors, this meant that they received little guidance from their superiors in the administration and top posts were turning over every few weeks. The entire administration, as one observer noted, "was virtually paralyzed by Watergate" at this point.[54]

In response to the appellate court's order, Reserve once again submitted its plan for an on-land tailings disposal site near Silver Bay. Judge Lord immediately rejected this plan, arguing that a tailings impoundment so close to the lake would inevitably leach tailings into the lake and that dust from the tailings would blow into Silver Bay. Judge Lord and federal prosecutors instead urged Reserve to move its entire grinding and separating operation to the mine at Babbitt. Reserve now claimed that if it was forced to relocate all operations to Babbitt it would shut down entirely. In August, the Eighth Circuit Court of Appeals again took up the Reserve case with a focus on potential settlement. Reserve now suggested a new tailings impoundment site near Silver Bay, known as the Lax Lake site. Suggesting Lax Lake as a tailings impoundment was a calculated decision by Reserve. The site had been suggested years earlier by pollution-control officials in Minnesota and it essentially forced Minnesota state officials to support the plan. After hearing Reserve's new plan, the appeals court ruled to stay Judge Lord's shutdown order indefinitely, effectively removing all pressure from Reserve to find an on-land site for tailings disposal. At this point, federal prosecutors again appealed the case to the Supreme Court. This time their appeal was joined by Solicitor General Robert Bork, who had received heavy pressure to support the case from environmental groups and members of Congress from the Great Lakes states. The Supreme Court again denied the appeal, but Justice William Douglas dissented, arguing that the court must balance the public's health with Reserve's right to profits. More important, the Supreme Court was impatient with the Eighth Circuit Court of Appeals's handling of the case and invited prosecutors to reapply to the Supreme Court if the case was not decided quickly.[55] The Eighth Circuit Court issued its final opinion in the appeal on March 14, 1975. It concluded that Reserve was polluting the air and water with asbestos-like fibers that likely posed some kind of threat to human health, although the exact nature of the threat was debatable given the scientific knowledge. The court also ruled that Reserve could remain open while it worked with the state to find a suitable on-land disposal site. Most important for the court, the ruling clearly stated that the pollution issues involved were outside the jurisdiction of the federal courts and that Judge Lord was not to take any further action on the case. Reserve Mining was now a matter for the state of Minnesota. Following the ruling of the Court of Appeals, the EPA quietly decided to quit pursuing the case. Officials who had worked for years to stop Reserve's pollution protested, but they were overruled by superiors in the Ford administration.[56]

From this point, Reserve Mining and state pollution-control officials battled over the location of an eventual on-land tailings basin. The decision to stop dumping into the lake was never in doubt, although haggling over the basin's location meant that more and more tailings were deposited into the lake until the basin was constructed. The government's argument, which was now being handled primary by pollution-control officials in the Minnesota Pollution Control Agency and the Department of Natural Resources, was that Reserve should relocate all operations to the mine in Babbitt. They were never satisfied with any dumping sites near Silver Bay. The insistence on dumping tailings in Babbitt also reflected the cultural attitudes and assumptions of environmentalists in the 1970s, many of whom loved Lake Superior and its North Shore as a recreational area but felt much less sympathy for the scarred and battered landscape of the Mesabi Range. Few environmental advocates asked whether additional pollution on the Iron Range would be equally problematic to pollution near Lake Superior.[57]

Throughout 1976, the state and Reserve battled over appropriate sites for on-land tailings disposal. Reaching an agreement was complicated by the distrust that had developed between state officials and representatives of the company. For example, when Reserve argued that any disposal site other than its preferred location was prohibitively expensive, the government noted that Reserve had said this about every site previously mentioned and had ultimately accepted new options. It was difficult to tell when the company was "crying wolf." By May, the two sides were debating between the so-called Mile Post 7 site outside Silver Bay, which Reserve preferred, and a more distant Mile Post 20 site preferred by state regulators. When the state denied Reserve the necessary permits to build the disposal basin at Mile Post 7, Reserve refused to accept the decision and enlisted public pressure from Silver Bay residents to push for the closer site. Silver Bay shops closed for one day in June so that residents could protest the state's decision at a meeting of DFL politicians in Duluth. Reserve also threatened to close its Silver Bay plant if it was not granted the preferred disposal site. Although some state regulators believed this was an idle threat, the courts took it seriously given its potential harm to the region's economy. Many of the environmentalists who were closely following the case were convinced that Reserve was bluffing in its threats to close the plant. They pointed out that Reserve could easily afford to spend millions on abatement and still make a healthy profit. The impasse over a dumping site was only resolved by the Minnesota Supreme Court, which ruled in April 1977 that the state was required to issue permits for Reserve to build the tailings impoundment at Mile Post 7.[58]

Despite predictions that pollution control would destroy Silver Bay's economy, the tailings impoundment proved to be an economic boon. Building the giant impoundment's dams and pumping systems employed almost three thousand construction workers. It was among the largest pollution abatement projects undertaken up to that time. Reserve reminded everyone that the $172 million it had spent to abate water and air pollution was "the largest single expenditure ever in the United States for environmental control purposes." The tailings impoundment was completed on March 16, 1980, and Reserve finally stopped dumping tailings into Lake Superior. More than a decade had passed since the state and federal governments first took action to stop the tailings disposal.[59]

Response on the Ground

While the trial occupied the nation's attention, residents of Silver Bay and Babbitt were faced with the more pressing issue of how to respond to claims of horrific pollution in their communities and the likelihood that they would lose their jobs if the plant was shut down. The small town of Silver Bay was hugely dependent on Reserve Mining. Silver Bay had been carved out of the North Shore forest two decades earlier expressly to house Reserve's workers and their families. By 1974, 95 percent of Silver Bay's workers were employees of Reserve Mining and the company paid more than half of the town's budget.[60] It was effectively a company town. Thus, the possibility that Reserve Mining could be shut down, bringing economic ruin to the town, weighed heavily on residents throughout the trial.

From the first moment that rumors of pollution surfaced, Reserve Mining sought to assure its employees and residents of Silver Bay that their jobs were secure and the town was not being polluted. Following the Stoddard Report's leak to newspapers in late 1969, Reserve Mining used the company newsletter to convince employees that the report was false. The newsletter told employees that the "erroneous 'report' was untrue and would be confirmed as

The small town of Silver Bay, Minnesota (left background of this photograph), *was built for workers at the E. W. Davis Works* (foreground). *Silver Bay residents were traumatized by the pollution lawsuit. They feared losing their jobs and worried about pollution harming their health. Photograph by Basgen Photography, circa 1965. Minnesota Historical Society Collection, HD3.27 p68.*

such in the near future." As the case developed, Reserve Mining and its parent companies expanded their public relations campaign nationwide. In April 1975, for example, Armco Steel chairman William Verity defended Reserve Mining and its environmental record in a letter to the *Chicago Tribune*: "If we had thought Reserve Mining's operations were a danger to health, we wouldn't have waited for a court order—we would have shut Reserve down ourselves." Reserve Mining Company and its executives maintained their innocence throughout the trial and sought to remind others that the facts about pollution were not settled.[61]

Reserve Mining's denial of the pollution charges illustrates how heavy industry, especially the steel and mining industries, responded to the environmental movement in the last third of the twentieth century. Once commodity producers recognized the extent of the environmental challenge, they organized into powerful alliances to dispute evidence of pollution and roll back new regulations. Organized business groups began what historian Samuel Hays called a "massive onslaught against environmental policies and programs" beginning in the 1970s. Industries banded together and turned to allied third parties, such as institutes and think tanks, to spread their message. Organized industrial groups argued that the environmental movement had gone too far, that meeting environmental regulations would eliminate jobs and hurt the economy, that the increased costs of meeting new regulations would be passed on to consumers and raise inflation, that industry should be allowed to self-regulate, and that well-meaning environmental regulations would have adverse consequences elsewhere in the environment. The steel industry, in particular, was shaken by environmental critiques. Steel companies had long basked in their position at the center of the economic health and well-being of the United States. But they also were several steps removed from the general public because they produced goods primarily for other industries. Thus, steel industry executives had little direct contact with the public and instead prized technological mastery over public relations. The Reserve Mining trial was an important event in pushing the steel industry toward an organized response to environmentalism.[62]

Silver Bay residents were shocked and surprised by Judge Lord's order to shut down the Reserve Mining plant in 1974. Although workers and their families closely followed the twists and turns of the trial, few had seriously considered that the plant would close and they would be thrown out of work. One welder at the plant told a reporter, "It's pretty hard to imagine that a huge industry would all of the sudden have the boom lowered . . . A lot of us figured there'd have to be a change in the disposal methods, but we didn't think the plant would be completely shut down."[63] Thus, Judge Lord's shutdown order in 1974 led to widespread confusion and panic. The union was flooded with calls from workers trying to determine whether they would receive unemployment benefits. When the plant was reopened by order of the Eighth Circuit Court of Appeals, workers in Silver Bay rejoiced. As news of the reopening spread through town, residents filled the streets of downtown. According to a police officer, that night downtown Silver Bay "sounded like pay day and Saturday night going to a barn dance." Silver Bay's mayor told reporters, "People are relieved that everything will be back to normal for at least a few more days," adding, "the people have a little more faith in the judicial system." Yet the whiplash of the plant's closing and reopening was only the beginning of several years of mounting anxiety over the plant's future in Silver Bay.[64]

Revelations that Reserve Mining threatened public health led to growing tensions between residents of Silver Bay, many of whom worked for Reserve Mining, and Duluth residents who felt as though they were risking their health to protect jobs in Silver Bay. At the beginning of the trial, many Duluth residents were sympathetic to Reserve Mining employees. In the small communities of the North Shore, many had relatives or friends who worked for the company. Yet the threat to Duluth's drinking water was more than many could take. Public opinion in Duluth quickly shifted toward criticism of the company and its pollution. Eleven thousand people signed a petition urging Reserve to stop dumping in the lake. A joke at the time in Duluth was that "it made little difference if Silver Bay residents went to hell; they wouldn't burn anyway." The pollution controversy divided the region along local lines, breaking down former solidarities.[65]

The prospect of harshly penalizing Reserve Mining over environmental pollution was spurred by a general boom in taconite production and employment during the late 1960s and throughout the 1970s. Unlike the 1950s and later in the 1980s, fears of unemployment and depression receded during the fifteen years between the taconite amendment's passage in 1964 and the onset of the steel crisis in the early 1980s (described in chapter 5). Thus, the Reserve Mining controversy took place against a backdrop of growth in the Iron Range's economic fortunes. Even as the rest of the nation sank into economic stagnation during the 1970s, the Iron Range prospered. Local boosters bragged that the region's economy was "moving well" and looking forward to a "permanent" expansion. Such optimism would ultimately prove to be shortsighted. The expansion of taconite in the late 1960s and 1970s was actually a brief respite between the earlier crises of the 1950s and the general collapse of the American steel industry in the early 1980s. But perceptions of growth and prosperity certainly made strict regulation much more acceptable than it would have been in other circumstances. The judges and prosecutors who confronted Reserve were under the impression that laid-off workers might quickly find new employment in the booming Iron Range economy.[66]

Other analyses painted a contrasting picture, suggesting that a full shutdown of Reserve Mining's mill would have devastating economic consequences for Silver Bay and the surrounding region. In 1973, the EPA conducted a study of the potential economic impact of closing the mill until tailings could be disposed on land. The study found that shutting down the E. W. Davis Works would eliminate three thousand jobs, result in $25 million in lost tax revenues, and lead to at least $4 million in state payments for unemployment. Another study funded by a pro-business group noted that employees laid off from Reserve would likely have a difficult time finding new jobs owing to their age and lack of transferable skills.[67]

The long-lasting environmental controversy also affected other parts of the iron and steel industry in the region. After watching the lawsuit unfold for several years, U.S. Steel decided to shutter its small Duluth steel mill in 1971. The mill had never been central to U.S. Steel's vast business empire, but environmental litigation, both at Reserve Mining and at the Duluth steel mill, were among the final straws pushing the company to shut down the mill.[68]

Despite the often hyperbolic alarm over jobs lost because of environmental regulations in the final decades of the twentieth century, a more evenhanded analysis reveals that increasing environmental regulations and antipollution sentiment created both winners and losers in the industrial economy. Sociologist Brian Obach argues that job loss resulting from environmental regulation was typically overstated by the media and industry. Although it is undeniable that certain industries lost jobs as a result of environmental regulation since the 1960s, Obach argues that these job losses were often smaller in number than alarmist predictions and offset by the growth of new industries connected to pollution abatement. Indeed, historians have found that pollution control often created as many jobs as it destroyed during the 1970s and 1980s. But there is no denying that the costs and benefits of these dislocations were not spread evenly. For older workers, many of whom had been with Reserve since its opening in the 1950s, a shutdown would permanently alter their life.[69]

The economic consequences of environmental regulations were also spread unevenly across the state's geography. Nationwide, enforcing pollution regulation often hardened a rural/urban divide over the environment. In most states, rural residents saw raw material industries such as mining and logging as sources of employment and they opposed environmental restrictions on those industries. In contrast, urban residents saw these extractive industries as a blight on the countryside—which many urban residents valued for its natural beauty and recreation value—and wanted the industries to be regulated more strictly. The Reserve Mining lawsuit, and the growing environmental movement in general, certainly divided rural and urban residents in the mining region. Residents of northeastern Minnesota's mining regions often resented environmentalists as out-of-towners who wanted to preserve the region as a wilderness playground and ignored their economic livelihood. One frustrated Iron Range resident attended a meeting of environmentalists with a large tree branch stuck in his shirt. Standing up, he yelled: "Look, I'm not a person, I'm a tree! Now will you pay attention to me?"[70]

On the ground in Silver Bay, residents were confused, scared, and angry over the trial, the looming prospect of a shutdown, and the silent threat of carcinogenic water. Following the whiplash of the plant's shutdown and reopening by the courts in 1974, residents lived in a constant limbo zone of possible layoff and dislocation. Living with this uncertainty had real costs on the health of Silver Bay residents. A schoolteacher noted that the constant threat of Reserve's shutdown during the trial deeply affected the town's schoolchildren. "They were anxious about losing their homes and where they would move," the teacher told a reporter. "They didn't care if their homework was done. Why should they do their schoolwork if they weren't going to be there?" Another teacher reported that incidents of vandalism and student drinking, smoking, and drug use in town were also rising.[71]

Labor unions at the national and local levels struggled to reconcile conflicting priorities throughout the Reserve Mining trial. Workers' jobs and paychecks were the primary focus for unions, as they had been for decades, but taconite workers' health and pollution in mining communities became an increasingly important issue as the trial progressed. Conflict within and among unions over environmentalism was not limited to the Iron Range. Indeed, labor's struggle to reconcile economic and environmental imperatives was a national problem during the postwar decades. Environmentalism's challenge to labor unions on the Iron Range was a microcosm of the challenge that an expanding environmental movement posed to labor unions throughout the United States during this era.

The fight between labor unions and the environmental movement was never as clear-cut as the "jobs versus the environment" rhetoric implied. As the primary representatives of working people, labor unions had a long history of working against industrial pollution and the deleterious health effects of heavy industry. The advent of large, integrated heavy industries in the United States brought with it an outbreak of industrial disease and sickness. Maiming, disease, and death were all too frequent in working-class communities. As historians David Rosner and Gerald Markowitz note, "health and safety was central to the lives of workers and to their organizing efforts throughout the twentieth century."[72]

Labor unions were also among the first institutions to sound the alarm over widespread industrial pollution and environmental degradation in the postwar era. Although labor unions were later portrayed as enemies of environmentalism, this was the result of developments that occurred in the 1970s and did not reflect the early postwar decades. During these decades, unions were among the most vigorous and powerful voices calling for environmental protection in the United States. Historian Scott Dewey argues that the nation's biggest unions, especially the United Automobile Workers (UAW) and the USWA, were "proto-environmentalists" that strongly championed environmental protection even when such positions were largely unpopular during the late 1940s and 1950s. Allied with Democratic politicians, the major unions were important proponents of antipollution legislation and the early legal framework of environmentalism. The unions did prioritize jobs and paychecks, but, as Dewey notes, unions "took [environmentalism] seriously, favored it consistently, and discussed it in terms and arguments sophisticated for the late 1950s and early 1960s," a time when few people in the United States were worried about such issues.

Tentative alliances between labor and environmentalists continued into the early 1970s before collapsing in the late 1970s and 1980s. Throughout the 1960s, a variant of labor environmentalism emerged throughout industrial America that promised, however tentatively, to connect labor's organized might to the emerging concerns of the environmental movement. Within the steel industry, labor began challenging steelmakers over pollution inside plants and in neighboring communities where workers and their families often dealt with the worst contamination from plants. Rank-and-file steelworkers started pushing for environmental regulations by the middle of the 1960s. In Pennsylvania's steel towns, USWA members were vocal advocates for clean air by the end of the 1960s. Several years passed before the USWA's leadership took up the pollution issue at the national level, but once they did the union became influential in passing the Clean Air Act and the Clean Water Act in the early 1970s. A district director for the USWA testified to Congress in 1970 that Chicago steelworkers wanted strict pollution laws applied to the steel industry, even if it meant that some union members would lose their jobs. Steelworkers in Peoria, Illinois, went on strike over the pollution emanating from the local steel mill, their employer. From the perspective of 1972, it appeared that labor environmentalism might become a powerful force for cleaning up America's land, air, and water.[73]

There were important exceptions to labor environmentalism during these years. Even at the height of its membership and political influence, the U.S. labor movement never spoke with a single voice. On the issue of pollution, the United Mine Workers of America (UMW) exemplified those unions that opposed environmental regulations. Throughout the 1960s and 1970s, the UMW leadership and rank and file opposed environmental regulations on mining, especially possible restrictions on strip mining. Although the Iron Range had been mined in large open pits since the turn of the century, other U.S. mining regions only developed strip mining after the Second World War when enormous earth-moving machinery could be adapted to new mining regions. The demolition of Appalachian coal-mining districts through strip mining attracted widespread attention—and condemnation—by the 1960s. The best-selling 1963 book *Night Comes to the Cumberlands* brought strip mining's ecological destruction to a wide and influential audience. In response, many environmentalists and politicians considered laws to ban strip mining altogether. The UMW organized against such restrictions, fearing that they would cripple an already damaged eastern coal-mining industry. Thus, the UMW took an active stand against new federal pollution-control laws in the early 1970s. Rather than support federal legislation, UMW officials supported regulation at the state level, where they presumed that it would be less strict. In the case of the mine workers, historian Chad Montrie argues that any professed environmental sentiments were "often insincere rhetoric to cover pragmatic economic concerns." Labor's stance toward the environmental movement was determined by local conditions rather than a nationwide approach.[74]

Like many other progressive alliances between labor and the New Left, however, labor environmentalism did not survive the economic turbulence of the late 1970s. Labor unions increasingly focused on bread-and-butter economic issues as American workers struggled with economic stagnation, inflation, the oil embargo, and other economic challenges. Within the environmental movement, a more radical element started to challenge the growth model that allowed both industrial workers and environmentalists to imagine a future of prosperity and clean air. Environmentalists who challenged economic growth itself had a difficult time finding common ground with labor unions. By the 1980s, labor unions and the environmental movement more or less parted ways. During the decade, unions focused on saving jobs from deindustrialization and the Reagan administration's open assault on labor. Environmentalists, for their part, took on several consumerist and conservative positions that turned off allies in the labor movement. Although there were some policies of common interest, such as right-to-know laws mandating that companies inform the public of hazardous chemicals used in manufacturing, these were the exception amid an overall trajectory of diverging paths.[75]

Tension between labor unions and environmentalists during the Reserve Mining trial was a microcosm of the broader national scene. Labor unions in the region at first supported strict antipollution legislation. When it later became clear that enforcement of pollution laws would affect employment in the taconite industry, unions allied with the mining companies against environmentalists in the name of job protection. The USWA, the main union representing both miners in Babbitt and workers at the E. W. Davis Works in Silver Bay, was deeply involved with the lawsuit from beginning to end. By the end of the trial, the USWA sided with the steel companies, choosing to prioritize jobs over potential environmental harm. Yet beneath the opposing public positions were a range of personal opinions that were often conflicted over the balance between economic security and environmental health.[76]

When Reserve Mining originally applied for permits in 1947 to dump tailings into Lake Superior, labor unions supported the plan. Like business groups and the local chambers of commerce, they focused on the jobs and payroll that the plant would bring to the depressed region. The only exception was the Brotherhood of Railroad Trainmen, which opposed the permits because it saw Reserve as a continuation of the earlier lumber companies that had clear-cut the northern forests. Amid the chorus of pro-growth voices, however, the Brotherhood's prescient opposition was ignored. Given the hyper-growth boosterism of the early postwar era, resistance to a potential new plant was ineffectual and, at times, politically damaging. Groups opposing the permits were labeled Communists.[77]

On other issues, especially outdoor recreation, northeastern Minnesota's labor unions were strong supporters of antipollution measures. In 1964, for example, the AFL–CIO urged state legislators to support environmental regulations. The union emphasized the need for outdoor recreational spaces for a newly affluent working class. There was, it wrote, "tremendous expansion of demand for outdoor recreation space so badly needed now and in the immediate future." For working Americans, especially blue-collar men in rural areas such as northeastern Minnesota, access to outdoor recreation such as hunting and fishing was a prime concern during the early postwar decades. Flush with newfound prosperity and union-backed vacations, workers wanted clean and healthy vacation areas for themselves and their families. Michigan autoworkers, for instance, treated fall deer hunting season as a sacrosanct working-class holiday. As historian Lisa Fine notes, "through the twentieth century, members of the working class came to understand their own access to the land and its game as both a right . . . and as an emblem of a particular type of working class masculinity." Labor's pre-1970s environmentalism drew from recreation and leisure pursuits more than workplace experience. When those two came into direct conflict, as they did during the trial, labor's response was far more ambiguous.[78]

As the trial developed and it became clear that Reserve Mining's Silver Bay plant might be shut down, labor unions in the region turned against environmentalism. Labor leaders in Minnesota argued that a handful of radical environmentalists and sensational press coverage had captured the public's attention and distorted views on the case. In a letter to other union officials in the region, the president of the ironworkers' union argued that the entire controversy was caused by sensational reporting by Twin Cities newspapers and environmentalists who lived outside the region: "Ordinarily, Twin Cities newspapers and citizens alike could care less about poor old Duluth, but in this case, wow!" The ironworkers' president then went further, embarking on a "one-man crusade" to convince politicians and business executives that allegations of Reserve's pollution were unfounded. He wrote to or visited dozens of politicians to present a packet of information demonstrating that Reserve was not polluting the lake and that the company brought invaluable jobs to the region. As the unions sided with Reserve Mining Company, environmental advocates increasingly saw unions as complicit in the pollution. When asked about the USWA's support for Reserve Mining, one environmental activist claimed, "labor has sold out for jobs, no matter what." It appeared that labor and environmentalists had little common ground as the case developed, truly pitting jobs against the environment.[79]

At an individual level, however, the line between worry over jobs and concern for the environment was never clear-cut. Many labor leaders expressed a genuine desire to clean up the environment and prevent pollution. Workers in Silver Bay, Babbitt, and Duluth were among those most directly affected by pollution, both on the job and at home. Gene Roach, president of the USWA in Silver Bay, was among those labor leaders deeply conflicted over Reserve's pollution. In a personal letter, Roach wrote, "It has always been our stand that if the tailings are causing an adverse effect on Lake Superior that we would have to take a stand against the discharge even if it meant giving up our jobs, homes and a good share of our lives. But we also felt Reserve should have opportunity for a fair trial based on facts." The case also began to divide the labor movement in the region. Those unions that directly benefited from Reserve Mining were generally supportive of the company, but unions not connected to the taconite industry worried about pollution. Labor leaders reported that the case pitted union against union depending on their relationship to—and benefit from—Reserve Mining.[80]

Few Silver Bay residents watched the unfolding drama with more sorrow than Edward W. Davis. After retiring from the University of Minnesota's Mines Experiment Station, Davis and his wife moved to a house in Silver Bay overlooking the giant plant named in his honor: the E. W. Davis Works. By all accounts, his years in Silver Bay were happy. He joked with reporters that he was the "undisputed patriarch" of the town and a local bar created a drink for him called the "taconite special." More substantively, he took great pride in his work with taconite, fondly looking over the plant—with its thousands of good-paying jobs—as recognition of his decades of hard work.[81]

Throughout his life, Davis had been an avid outdoorsman. He and his wife frequently spent long summers in the north woods, where he wrote loving descriptions of his time fishing and canoeing amid the clear, cool waters of northern Minnesota lakes.[82] Although Reserve Mining's opponents would later portray Davis as the chief engineer of a great system of pollution, this description fails to capture his complex mixture of appreciation for nature and ignorance of the consequences of his actions as a mining engineer. For as much as Davis enjoyed the natural world in his personal life, his professional opinions revealed an ethos of development that imagined the natural world as a storehouse of resources and ignored obvious ecological consequences from large-scale mining operations. For example, he eagerly imagined how atomic weapons could be used in mining operations. He urged Minnesota to subsidize iron-ore exploration at great depths by using atomic blasting to strip surface rock. He imagined future mining operations, controlled by the Atomic Energy Commission, that would blow up much of the Mesabi Range to reveal valuable taconite deep below the surface. He also proposed that the gaping holes left by open-pit mines could be filled with junk cars once they were no longer used for mining and then blended with water and chemicals to promote oxidation.[83] Davis's personal contradictions regarding the environment—enjoying untrammeled wilderness in his personal life while advocating deeply destructive mining practices as a professional—were typical among mining engineers and metallurgists of his generation. As mining historian Duane Smith notes, mining engineers throughout the United States often described the wilderness as a site for character building, but rarely considered how that same environment would be affected by mining operations. Davis thus typified the development-minded engineer of the mid-twentieth century: he enjoyed spending time in the woods but was never especially concerned about pollution.[84]

Concern over pollution from the E. W. Davis Works stunned Davis. Although he was in his eighties and had been retired for many years, he once again tried to influence public opinion by writing to state officials and newspaper editors. In a 1972 letter to the Minnesota Pollution Control Agency, he argued that it was inaccurate to say that tailings were being "dumped" into the lake because, in his opinion, most of the course tailings remained in a delta near the shore and only the finer particles made their way into the lake:

> As a resident of Silver Bay and a user of the water in this area, I am strongly of the opinion that Reserve's taconite tailings are being disposed of at the present time in the best possible manner. If the tailings were being deposited elsewhere on land they would have to be contended with for years to come. As it is now,

when the taconite operation is finished there will be nothing to show that it ever operated in the area.

Davis also expressed his growing exasperation with the public's changing perception of industrial waste and was incredulous at suggestions that taconite tailings were carcinogenic. "There is no asbestos in it!" he argued. "It's all supposition that this material will have the same effect as asbestos. It's all hypothetical . . . we've been drinking this water for years!"[85]

In 1973, with his health failing, Davis once again reached out to his political connections in a final attempt to rescue taconite from the taint of pollution. He urged his daughter to contact his old friend, Representative Blatnik, regarding the Reserve Mining case. Davis apparently believed that Blatnik would somehow be able to influence the case against Reserve. In the letter, Davis's daughter told Blatnik that her father "either doesn't understand or doesn't want to" that the pollution case was going forward. Poignantly, his daughter wrote to Blatnik: "I feel sorry for him too . . . it was such a nice life to have given taconite to the U.S. He doesn't want to have polluted Lake Superior; he loves it. I wonder sometimes . . . how fair we are to damn the log-cabin builders who chopped down the forests."[86] Davis died later that year.

Blatnik, usually placid in his correspondence, replied to the letter from Davis's daughter with surprising emotion:

> Many years have passed since we joined together, he in the laboratory, I in the State Senate, to make the dream of taconite reality. Those years have made a substantial improvement in the living standards of everyone in Northeastern Minnesota; and have contributed materially to the industrial might of the United States. Whatever the outcome of these present very difficult circumstances, your father's contributions cannot be denied him. History will surely preserve his role as the single most important individual in keeping the Iron Range alive. But I, like you, grieve that he should have to see his lifework, his foresight and great scientific abilities questioned, however briefly. Let us pray for a swift vindication of one of the truly great men of Minnesota.[87]

Blatnik was sad that his friend's final years were spent worrying that his great technological accomplishment might be polluting

Lake Superior and poisoning the people who lived near the lake. Implicitly, Blatnik's letter also expressed his sorrow for a passing age, when the technological fix could be celebrated with little or no regard for its costs to the environment. By the 1970s, environmental factors had forced themselves back onto the ledger sheet.

Conclusion

"It is safe to say that everyone is weary of the Reserve Mining controversy," wrote Robert Bartlett in his 1980 analysis of the trial. After more than a decade of administrative hearings, trials, and scientific studies, the controversy left few people satisfied with the outcome.[88]

Once the asbestos issue took center stage in the trial, the question of whether Reserve was polluting Lake Superior fell by the wayside. This question was never answered definitively. By the time Reserve Mining switched to on-land tailings disposal in 1980, many of the initial questions about "green water" and tailings' effect on aquatic life were less pressing. Judges at the state and federal level had issued contradictory rulings on whether the tailings did indeed constitute pollution and the mountain of scientific data produced by the trials only led to more confusion. Looking back, one observer noted, "after detailed testimony from such an impressive array of MDs, engineers, multi-disciplined scientists, economists, and others, facts in the case presumably would be established and the final verdict would be clear-cut. Actually the reverse happened."[89] For environmentalists and their opponents, the bruising courtroom battle left lingering questions of whether environmentalism could be enforced—or rejected—by the courts and how society as a whole could possibly balance such sharply contrasting values.

The asbestos-like fibers in Lake Superior did not prove to be deadly. Testing conducted since the 1970s has found little evidence that taconite tailings produced higher-than-expected rates of cancer in the region or even among taconite workers. This is not to say that taconite had been proven entirely safe. When the University of Minnesota tried to convert the former Mines Experiment Station building into student dorms in 1997, the plan had to be scrapped because of much higher than expected environmental cleanup costs. The building contained asbestos and decades' worth of hazardous chemicals. A connection between asbestos and taconite dust was revived in 2006 when the Minnesota Department of Health admitted that thirty-five Iron Range miners had developed a rare and deadly form of cancer often related to asbestos dust. Subsequent testing, as of 2013, has been inconclusive.[90] Yet, in the final analysis, it is undeniable that the public health risks from taconite were wildly overblown during the trial. Even Grant Merritt, one of Reserve Mining's sharpest critics, later admitted, "I haven't seen a study that shows absolute proof or solid scientific evidence that ingestion of asbestos-like foreign fibers will cause mesothelioma."[91] Since the early 1970s, Americans have perhaps come to accept that modern life is filled with carcinogenic risks that cannot be fully eliminated. Also, eliminating stubborn environmental pollution has proven more difficult than focusing on cancer risks that can be controlled by individual behavior, such as smoking.

The trial, and the larger issue of industrial pollution, forced miners, engineers, and other proponents of economic development on the Iron Range to confront new and complex factors when they made decisions about future development. Beginning in the 1970s, questions of how economic development might affect the environment had to be factored into decisions alongside job numbers. In some cases, the environmental harm of a new mine or factory was so great that it was not worth it, even if it produced jobs and tax revenues. Indeed, weighing environmental costs versus economic benefits became the central question when new mining projects were proposed after the 1970s. While this may seem logical from a twenty-first-century perspective, it is worth recalling the challenge that this posed to the older ethos of economic development. Reserve Mining's many defenders upheld this older system in the face of new demands. In Robert Bartlett's words, Reserve Mining's defenders saw themselves as "believing in, and defending, values that had once been almost universal." It was no small thing to ask them to abandon a credo of technology-led economic development in favor of the much more nebulous demand for ecological balance. A reporter covering the trial captured the sense of loss felt by many of the Iron Range's old guard: "back in 1958 it all seemed so simple. Jobs, iron, prosperity, and no complications. If we are wiser now, we are sadder as well." The heroic era of economic development— led by brave politicians such as John Blatnik and wise scientists like Edward W. Davis—was over, brought down by the unintended consequences of its own success.[92]

In the broadest perspective, the Reserve Mining case raised difficult questions about the ultimate relationship between industry and the environment. The entire history of the Iron Range had taken place within a global history of modernization, economic growth, and technological development. Indeed, tapping into the iron ore provided by the Iron Range's geology had been central to the twentieth century's tremendous industrial growth and development, with Lake Superior ores providing an essential raw material for the rapidly expanding steel industry. When that system proved to be unsustainable owing to depletion of the high-grade ores, taconite was created as a grand technological fix. Yet the technological fix of taconite was only effective because the costs of low-grade iron-ore mining and processing were shifted onto the environment. When those environmental costs were revealed in the Reserve Mining trial, the fix appeared far less painless. Low-grade iron ore thus offers a potent example of the modernist belief that the latest science and technology could obliterate all natural limits, including the finite deposits of high-grade iron ore, and the larger mind-set that fundamentally separated nature from technology.

Finally, many of the concerns over ongoing pollution were made moot by the steel crisis of the 1980s. As described in chapter 5, the downturn that swept through the American economy in the early 1980s decimated the steel industry and shuttered several of the Iron Range's mines. Reserve Mining Company was among the losers. The company went bankrupt and closed all operations in 1986. Most of Reserve Mining's workers lost both their jobs and their pensions. The plant at Silver Bay sat empty for several years, until it was reopened in 1989 as North Shore Mining, a small, nonunion plant run by Cleveland-Cliffs.[93]

FOUR

Economic Development Policy, Regional and Local
The Iron Range Resources and Rehabilitation Board

Chapters 1 through 3 described the most significant responses to industrial decline on the Iron Range during the twentieth century. First, taconite was invented as a technological fix for the declining natural iron-ore mining industry. Through the miracles of science and engineering, Edward W. Davis and his colleagues created an industrial system that promised a future of ongoing mining and abundance. Yet taconite technology alone did not spur the Iron Range's mining economy. It needed assistance from politicians in the form of revised tax laws. The 1964 taconite amendment revised Minnesota's tax laws to make taconite commercially viable, but it also heralded a new era of tax-cut politics that hollowed out the labor–liberal alliance. Finally, the Reserve Mining pollution saga revealed the hidden environmental costs of the technological fix. What first appeared as a painless solution for decline had, in fact, displaced those costs onto the environment. In its birth and

133

complicated life, taconite highlighted the difficulty of saving an industrial region through new technology. Yet efforts to create and sustain a taconite industry were not the only attempts to prevent economic decline on the Iron Range in the years after World War II. This chapter focuses on economic development policy as another important way in which the Iron Range fought against economic decline in the twentieth century.

Growing government involvement in the economic welfare of everyday Americans was one of the most significant developments in the United States during the twentieth century. Beginning in the Progressive Era and expanding greatly during the New Deal and World War II, governments at the federal, state, and local levels took it upon themselves to protect the economic security and well-being of those living within their borders. The government took on the responsibility of providing citizens what President Franklin Roosevelt called "security against the hazards and vicissitudes of life." The Social Security Act of 1935 was only the most prominent example of a growing American welfare state that took citizens' economic security as a primary concern. Although it was always greatly contested, this governmental commitment to economic security and welfare continued into the postwar years. During the Great Society of the 1960s, for example, President Lyndon Johnson committed the federal government to nothing less than eradicating poverty and ensuring that all Americans had access to postwar prosperity, asking Americans to "join in the battle to give every citizen an escape from the crushing weight of poverty."[1]

Beneath the soaring presidential rhetoric, the federal government relied on broad-brush strategies to help the economy during periodic downturns. Although there were brief experiments with direct regional planning by the federal government during the New Deal, the United States mainly relied on macroeconomic levers to support the national economy as a whole throughout the postwar years. It was the era of push-button Keynesianism. Unlike governments in Western Europe that enacted national industrial policies to coordinate capital and labor in industries deemed vital to the nation's economic health, the United States never pursued a coordinated industrial policy in the postwar decades. American economic policy during the middle of the twentieth century therefore contained profound contradictions. At the federal level, the government was simultaneously committed to ensuring the economic welfare of its citizens but hesitant to enact policies that would do so in regions such as the Iron Range that were suffering from industrial decline.[2]

Economic development policy was a key policy response to declining local economies in the postwar decades. Historians have described numerous local economic development efforts in the United States at midcentury. For example, in cities hit by early waves of deindustrialization, such as Detroit, Philadelphia, and St. Louis, city and state governments turned to economic development policy early in the twentieth century. Rural industrial regions similar to the Iron Range also adopted economic development policies during the postwar era. These myriad local, regional, and state policies formed what historian Otis Graham Jr. calls a "de facto industrial policy" pursued in the United States. The problem, though, was that this tangle of local policies often worked at cross-purposes. Nor was economic development policy limited to the United States during the twentieth century. Indeed, there was a transnational consensus on economic development policy and developmental economics in the postwar decades. Around the globe, international institutions such as the World Bank, the International Monetary Fund, and the United States Agency for International Development promoted economic development measures to stimulate the economies of poor nations. From city governments to international agencies, economic development was one of the main policy responses to economic decline and stagnation in the postwar era.[3]

On Minnesota's Iron Range, the primary government agency charged with coordinating economic development policy was the Iron Range Resources and Rehabilitation Board (IRRRB). This chapter describes the IRRRB and its history to illuminate the institutional context in which responses to industrial decline occurred. The IRRRB's history not only illuminates the little-known history of state and local economic development policy in response to deindustrialization, but also illustrates the limits imposed on postwar economic development policy by relying on local governments to respond to what were ultimately global problems of capital flight and deindustrialization.[4]

Historians of public policy and American political development have mapped the consequences that sprang from particular institutional choices in American politics. For instance, historian Julian Zelizer has demonstrated how Congress's committee structure allowed liberals to defend activist policies and taxation against widespread public hostility during the 1950s and 1960s. Applying a similar analysis to economic development policy raises important questions about the range of possibilities available to policy makers who confronted economic decline in the twentieth century. How did the local focus of postwar economic development policy constrain the options available to policy makers? Were there possibilities at the state and local levels that were not available had economic development been pursued by the federal government?[5]

Origins of the IRRRB

The Iron Range Resources and Rehabilitation Board was born out of the economic dislocations of the Great Depression and the turbulent politics of the New Deal. Like many industrial regions of the United States, the Iron Range's economy was decimated by the Great Depression. Iron Rangers faced high unemployment and growing misery as the Depression wore on. Observers who traveled through the Iron Range during the 1930s noted that unemployment resulted not just from temporary economic conditions, but also from deep structural factors, including mechanization and automation in the mines. Reporter Lorena Hickok visited the Iron Range in 1933. Writing to Eleanor Roosevelt, Hickok reported that the Iron Range had "an unemployment problem . . . that will never be cured, probably." She saw that automation, not the Depression, was driving workers out of the mines.[6]

The shutdown of so many of the region's mines during the Depression forced civic leaders to consider the possibility that iron-ore mining might not have a long-term future in the area. Perhaps this realization should not have been surprising given the inherently unsustainable nature of a one-industry economy based on resource extraction. But some municipal leaders on the Iron Range took away a different lesson: they were determined to save their communities and make sure they did not dissolve into a string of ghost towns. Instead of succumbing to capitalism's creative destruction—the inherent impermanence of a mining boomtown—the Iron Range's civic leaders believed that the region and its mining economy should be preserved for the future. From the very beginning, then, the premise animating economic development policy on the Iron Range was a refusal to accept decline as inherent to resource-extraction industries and their communities. Economic development policy on the Iron Range, as embodied by the IRRRB, was always guided by the principle that the impermanent could be made stable and permanent.[7]

Yet the formal shape of an economic development agency was initially unclear. Civic leaders concerned about the Iron Range's economy first pursued development funds from the federal government. In May 1941, Iron Range banker Fred Cina, who would go on to a distinguished career as a state legislator, typed a long letter to President Franklin Roosevelt explaining the dire employment situation on the Iron Range and urging Roosevelt to provide federal aid to the region. Cina emphasized the Iron Range's importance to the U.S. steel industry by arguing that the Iron Range's mines "produce 75% of the total iron ore in the United States." More important, he argued that increased production for defense had not relieved the Iron Range's unemployment problem. Cina noted that improved mining technology in the open-pit mines allowed ore to be removed with fewer and fewer human workers. "The result of this increased output per man," he wrote, "has been to curtail the employment of the men in the iron region, (mining being the sole industry) so that thousands have been forced onto the relief rolls of the state, and large numbers have been provided for by the W.P.A. [Works Progress Administration]." The letter painted a dark portrait of a region where "the population of some 70 thousand is entirely stranded, and will be, unless a rehabilitation program is carried out by the nation."[8]

Cina's call for federal aid was received lukewarmly in the White House and on Capitol Hill. His letter was forwarded to several congressmen, who replied that they understood the need for relief on the Iron Range but that national defense now took precedence over economic relief. President Roosevelt was supportive but noncommittal. As the federal government shifted from domestic to international concerns in 1941—as Doctor New Deal gave way to Doctor Win the War—the federal government stepped away from its previous concern with declining industrial regions.[9]

With little prospect of help from the federal government, efforts to create an agency for economic development on the Iron Range focused on the state level. The immediate impetus for creating a state agency for Iron Range development was the 1941 change in Minnesota's mineral tax laws described in chapter 1. In addition to eliminating the ad valorem tax on taconite, Harold Stassen, Minnesota's governor at the time, pushed for other changes in the

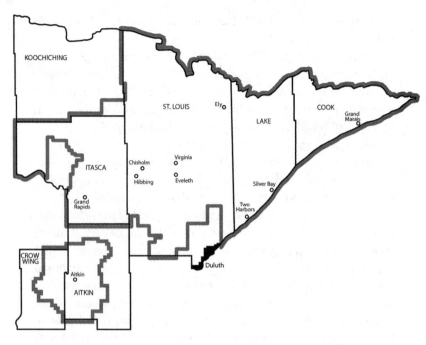

The Taconite Assistance Area, the region served by the Iron Range Resources and Rehabilitation Board. Courtesy of Iron Range Resources.

state's tax code. These changes included limiting local tax levies to $70 per capita and $60 per capita for school districts, changes that were aimed at what was widely considered to be profligate municipal spending by Iron Range communities. The law also increased the occupation tax, which taxed the actual amount of iron ore mined, and a program to funnel a portion of the mining occupation tax into a "rehabilitation program" for the Iron Range. The last of these changes, the rehabilitation program funded by a portion of the iron-ore occupation tax, was the germ for the IRRRB.[10]

The agency's initial mission was to oversee the pool of occupation tax money and direct it toward appropriate rehabilitation activities. Originally named the Department of Iron Range Resources and Rehabilitation, the agency was later renamed the Iron Range Resources and Rehabilitation Board. The board represented a cross section of the Iron Range's political interests; it was composed of a shifting group of Iron Range state senators and representatives. The IRRRB's primary focus was the mining region and it spent the majority of its time and money in the early years on the iron-ore mining industry. But the agency was also charged with developing the Iron Range's "other resources, both human and natural." As this mandate was understood in the 1940s, developing other resources meant creating a new wood products industry, encouraging specialized agriculture, building educational facilities, and promoting tourism in the region. In short, the agency saw its task as diversifying the Iron Range's economy away from its singular dependence on iron-ore mining.[11]

Despite this broad mission, the agency's original financial situation was precarious. By tying the agency's funding to the iron-ore occupation tax, the 1941 law created a funding problem. The IRRRB was trying to act as a countercyclical government agency, spending government money during downturns and saving during booms in the iron-ore mining industry. Yet the IRRRB's revenues followed the ups and downs of the mining industry because the occupation tax was imposed on the amount of ore mined in any given year. There was also no provision in the original law for the agency to save funds for later use. When the iron-ore mining industry was flush, the agency's income was high but there was less demand for its services. When the industry was down, however, Iron Range communities demanded assistance from the IRRRB, yet there was little money available from tax revenues. The agency's initial funding structure made it difficult to achieve its mission of buffering economic downturns in the region.[12]

The IRRRB in the 1950s and 1960s

The IRRRB initially struggled to define its place in Minnesota's postwar political landscape. The agency was conceived as a response to the Great Depression, but its continued existence into the 1950s forced the agency to experiment with a variety of new development projects. During these years the IRRRB also faced numerous critics who wanted to dissolve the agency altogether. Although the IRRRB's ultimate direction remained unclear for years, during the 1950s the agency established its foundation as a state-level, public–private economic development agency controlling mining tax revenue.

Many of the IRRRB's early problems stemmed from a lack of clarity in the statutes that created the agency. Originally established in response to the Iron Range's high unemployment rates during the Great Depression, the agency's duties were unclear amid the changed economic landscape of the 1950s. The Iron Range economy was struggling throughout the decade—as described in chapter 2—but the crisis atmosphere of the nationwide depression was gone. In response, some Minnesota politicians argued that the easiest solution would be to dissolve the IRRRB entirely. This solution proved politically unfeasible. Few politicians in Minnesota were willing to alienate the Iron Range's powerful voting blocs. But politicians nonetheless did their best to limit the IRRRB's power and influence by cutting its funding or diverting funds away from the IRRRB and toward other state agencies. The IRRRB received significant funds from the iron-ore occupation tax throughout the 1950s. Yet a sizable portion of those funds were transferred out of the IRRRB and put toward other state agencies. The IRRRB's total income from the occupation tax was therefore always somewhat misleading. The agency had less money to fund rehabilitation than what appeared on its balance sheets. Even if Minnesota politics made it impossible to eliminate the agency, some politicians tried to reduce the IRRRB to an ineffectual agency that simply passed tax revenue on to other state agencies.[13]

Another factor contributing to ambivalence toward the IRRRB was an underlying belief among some state politicians that economic distress was endemic to the Iron Range and government action by the IRRRB would only prolong the suffering. For example, a 1940s report on northeastern Minnesota's industrial problems revealed that some state officials thought wholesale out-migration from the region would be the best solution to the Iron Range's economic malaise. If there was no work in the area, the thinking went, then people should just leave. As the report described it, migration out of the Iron Range "is probably the simplest solution to the Mesabi Range problem." Although not all policy makers agreed with this view, it reveals the deep ambivalence about the underlying needs of the Iron Range and the mission of the IRRRB. If the only viable long-term solution to the Iron Range's economic problems was migration out of the area, why should the state government commit to wholesale redevelopment?[14]

Indeed, opinions about whether or not the Iron Range's economy needed government-sponsored redevelopment in the 1950s were colored by national debates over depressed areas amid the postwar economic boom. While the nation as a whole was enjoying unprecedented growth and abundance, some pockets of the nation—such as Appalachia, inner cities, and the Iron Range—were clearly being left behind. There was very little consensus about how, or whether, such regions needed government assistance. Advocates of economic development policy looked back to the planning traditions established during the New Deal and argued for vigorous public policy responses that would bring these regions up to the national standard. But more conservative voices noted that the nation's uneven economic growth was the inevitable result of capitalism and technological change. The winds of "creative destruction" blew across the nation's economy, conservatives argued, and it would be foolhardy to think that government policy could arrest seemingly inevitable processes of economic change and, for some areas, decline.[15]

Supporters of the IRRRB, in contrast, argued that a regional development agency had a valuable role to play in Minnesota's postwar economy. IRRRB commissioner Edward Bayuk admitted in 1952 that increased iron-ore production had lifted the region's most acute employment problems during the early 1950s. But he nonetheless argued that the state should continue working "with an eye to the future" to create a "better balanced economy as cushion for the future as well as to meet the immediate needs of . . . new vocational fields." The IRRRB, he said, offered a unique opportunity to continue the planning tradition of the 1930s and 1940s in the postwar decades. Prominent Minnesota businessman Charles L. Horn agreed with Bayuk, publicly stating that the IRRRB should not be dissolved and could be a valuable contributor to Minnesota's long-term growth. In a 1951 memo, Horn wrote that the problems facing the Iron Range were "not just local or sectional" but "state and national problems" and only a dedicated agency such as the IRRRB could coordinate economic development efforts involving government, business, and academic experts.[16]

Political controversy surrounding the IRRRB was exacerbated when the Minnesota attorney general's office circumscribed the scope of economic development projects that the agency could legally undertake. In the middle of the 1950s, the attorney general's office sharply limited the type of projects the IRRRB could participate in, determining that "these projects must be something new or in the form of research." In other words, the IRRRB could not funnel public money into existing industries even if they were struggling. As a result, the IRRRB turned down many requests for development projects throughout the decade. The agency's ability to own or rent land came under special scrutiny from the attorney general. Previously, the IRRRB set up pilot plants for industrial and agricultural experiments on publicly owned lands. This was deemed to be too close to governmental competition with private industry. Eventually, a special assistant attorney general was attached to the IRRRB to rule on whether individual projects were within the agency's legal scope. In a broad sense, the limits on the IRRRB's activities illustrated widespread hesitation about activist governmental involvement in the private economy in the 1950s.[17]

Despite these questions about the IRRRB's existence and mission, the agency nonetheless pushed ahead on a variety of economic development projects during this period. In its 1950 report to the Minnesota legislature, the IRRRB highlighted a range of economic development projects it had supported, including aerial photography of the Iron Range, cow testing, forest surveys, a failed iron powder project, peat research, potato farming, and titanium drilling. The IRRRB, which was struggling to define its mission, paid for many different research projects in the hopes that they would diversify the region's economy.[18]

Projects to support the logging industry were one focus of the IRRRB's early efforts. Because the area covered by the IRRRB's jurisdiction included forested areas, the agency emphasized development of the logging and wood products industry as a useful way of diversifying the region's economy. Logging had a long history in the upper Great Lakes region. In fact, logging predated the iron-ore mining industry in the region by several decades. Yet the Iron Range's old-growth trees were logged off by the end of the nineteenth century and the industry contracted sharply. The resulting wasteland of stumps and brush was commonly referred to as the "cutover" area.[19]

To stimulate the timber industry in the cutover area, the IRRRB joined other state agencies to propose several new programs aimed at economic development through improved management of the region's timber resources. For instance, Iron Range civic officials initially proposed that that region create community forests. Worried about declining employment in the iron-ore mining industry, these city leaders thought that selective logging in community-owned forests could employ people in a resource-extraction industry that would rejuvenate itself, unlike iron ore, and would offer sustainable employment. The community forests were never implemented, but they illustrate the imaginative possibilities for economic development being discussed in the 1940s and 1950s. The IRRRB was especially interested in developing paper mills because they could potentially offer high-wage industrial employment and would complement northeastern Minnesota's overall resource-extraction industrial base. Unfortunately, programs to support the timber industry were lower on the IRRRB's list of priorities than mining and they were among the first to be cut when the agency's budgets dropped in the 1950s and 1960s. Throughout the 1960s, the agency scaled back its logging programs because of declining revenue.[20]

Among the more creative economic development projects pursued by the IRRRB during its early years was support for Jeno Paulucci, a Duluth entrepreneur. Paulucci, who was born on the Iron Range, pioneered the making and marketing of prepared ethnic and frozen foods during the 1950s. Paulucci approached the IRRRB and asked for several hundred thousand dollars to expand his Duluth-based Chun King Asian food business. More important, he threatened to leave Minnesota, taking his company's jobs with him, if the IRRRB did not offer him a loan. The agency noted that Paulucci "was even considering leaving the state. He needed more space and one state already had offered him the sort of plant he wanted." The IRRRB offered him a loan of two hundred thousand dollars, which, the agency later bragged, was "paying off handsomely due to increases in farming and payroll in the Duluth area." Although the loan to Chun King paid off, Paulucci's threat to leave the state was a harbinger of the problem of interstate capital flight and the susceptibility of state-based economic development agencies to threats that the plant would be moved out of the state. Once committed to spending public funds to keep jobs in a given region, economic development agencies could become pawns in business owners' demands for public financing. By threatening to leave an area for a more lucrative deal, owners could, at times, blackmail agencies into ever larger pools of public money. Paulucci likely understood this well. He was a savvy political operator who went on to finance Hubert H. Humphrey's failed 1968 presidential campaign and was in touch with many postwar presidents. The problem of competing economic development agencies was not limited to the Iron Range. Because states and localities, rather than the federal government, took the lead in postwar economic development policy, there was often sharp competition among regions for industrial jobs. Cities and states promised tax cuts and other incentives to lure new industries or retain existing ones. The result was often an economic development arms race between regions as they promised more and more public support for corporate wishes.[21]

During the first two decades of its existence, then, the IRRRB struggled to clearly define its mission. It moved simultaneously in several different directions, while struggling with growing restrictions on its activities by a state government increasingly skeptical of direct government intervention in the economy. Ultimately, it was not until the arrival of broad federal support for regional development agencies in the 1960s that the IRRRB's place within state government was assured.

IRRRB and Area Redevelopment

In the early 1960s, the IRRRB's problems were partially offset by the creation of a new federal agency charged with regional economic development: the Area Redevelopment Administration (ARA). A short-lived federal program, the ARA was a tentative and ultimately failed step toward a coordinated national policy response to industrial decline and regional deindustrialization in the middle of the twentieth century. For a few years during the 1960s, however, the IRRRB worked closely with federal officials in the ARA to coordinate economic development aimed at improving the economy of the Iron Range.

The ARA was the brainchild of Senator Paul Douglas of Illinois. After touring portions of his home state during the mid-1950s, he became convinced that some sections of the nation were not partaking in the postwar economic boom and believed that the federal government should ensure that all regions shared in the era's economic prosperity. "Douglas gained the conviction," one associate wrote, "that the federal government has the responsibility for assisting depressed areas in rehabilitating themselves." Specifically, Douglas thought that depressed areas did not have enough "indigenous resources" to create their own prosperity and thus needed "specially tailored programs."[22] Douglas was also influenced by industrial policies in Western Europe that framed deindustrialization as a national problem in the postwar decades. These beliefs were translated into policy in 1955 when he first introduced the Depressed Areas Act, a bill for more than $100 million in federal funds to create public facilities, train workers, and promote businesses in depressed regions. Speaking in defense of his bill, Senator Douglas said, "It is inhuman to let these areas rot away. The lives of too many human beings are at stake to sit by and do nothing for these pockets of depressed industries and localities while much of the

rest of the country enjoys a high standard of life."[23] Douglas's bill
was quickly caught up in partisan politics, however, with congressional liberals debating the Eisenhower administration over how
much money was needed and where the administration would fit
within the federal government. Eisenhower eventually vetoed two
versions of Douglas's bill, which had expanded to cover both urban
and rural poverty in nearly all areas of the country. As it moved
through Congress, the depressed areas legislation grew in ambition
while being cut down in size and funds. Historian Judith Stein, for
example, is critical of the program's assistance to the steel industry, noting that proposed legislation offered only "small sums" that
were "thinly spread" across the country, meaning that little concentrated development in a given industry or region could occur
with the federal money. Historian Gregory Wilson concludes that
the ARA did not "adequately address the issue of distressed communities, leaving these labor markets with a weak cushion against
the ill effects of industrial transformation."[24]

John F. Kennedy seized on depressed areas as a political gam-
bit during his 1960 presidential campaign. Kennedy had worked on
the act while in the Senate Committee on Labor and Public Welfare. More important, his home state of Massachusetts was beset
by early waves of industrial decline, particularly in former textile
manufacturing towns such as Lowell, and he was thus sensitive to
the problems facing declining regions. During the 1960 election,
the problem of depressed regions came to the fore in West Virginia,
where the primary campaign hinged on poverty in the mountain-
ous mining region. After winning the election, Kennedy made the
Area Redevelopment Act an early priority for his administration
and on May 1, 1961, he signed it into law. At the core of the ARA were
four key policy initiatives: low-interest loans for expanding businesses in depressed areas, direct federal aid to depressed communities for building public facilities that would help to attract business, training programs for the unemployed, and planning help for
communities hoping to attract new business. While helpful, these
programs were a far cry from the more robust federal role imagined
by Paul Douglas, who had initially hoped to revive the planning
traditions of the New Deal. The ARA's problem was that the political landscape of 1960 was a far cry from the 1930s. The overarch-
ing political context under which the ARA operated was the stifling
anticommunism of the Cold War, which meant that redevelopment
policies from the federal to the local level had to function within

the market economy. Any proposals hinting at direct government involvement in industry were suspect. As a result, the ARA and local economic development programs relied on job training, loans to developers, or limited public works projects rather than wider-ranging interventions in the free market.[25]

The ARA also drew heavily on state-level agencies as a model for a federal economic development agency. "Federal redevelopment and employment policy," Wilson observes, "moved along paths already worn by local and state governments." Historians have pointed to other regional economic development plans, notably a Pennsylvania agency and organizations tackling poverty in the New England textile region, as state-level models for the ARA. But Minnesota's IRRRB should also be included in the list of regional economic development agencies that influenced the creation of the ARA.[26]

President Kennedy had a close connection to the Iron Range through his friend, Representative John Blatnik from Minnesota's Eighth Congressional District. Blatnik and Kennedy both entered Congress as new Democratic legislators in 1946. They were both veterans of World War II and they worked together closely during their early years in Congress. Blatnik noted that he and Kennedy were two of the only "young fellows" in the House of Representatives at the time and "we became very close friends." More substantively, Blatnik influenced Kennedy's thinking on the problems of poverty in the United States. While Kennedy had a good understanding of urban poverty, he relied on Blatnik to help him understand the question of resource usage in the United States and the connection between resource extraction and poverty. Recalling his conversations with Kennedy on the matter of resource use, Blatnik said:

> we discussed in detail that more attention ought to be paid to these areas that we now call distressed areas . . . There was real severe economic distress in these areas which affected millions of people; it was easily one fifth of the total population of the country, and much of it revolved around the resources that had been exploited or poorly used or maybe exhausted. There ought to be serious concentrated attention paid to rebuild those areas, and many of them could be rebuilt.

Blatnik and Kennedy were also close political allies. During the 1960 presidential campaign, Blatnik organized a wildly popular campaign stop for Kennedy on the Iron Range. He filled the Hibbing gymnasium to the rafters with supporters. He later argued that Kennedy's support from the Iron Range likely swung Minnesota to his side during the 1960 election. Thus, it was not surprising that the Iron Range was on Kennedy's mind as a blighted region when he drew up the ARA.[27]

Immediately after the ARA was created, the IRRRB began working with the federal agency to coordinate economic development efforts in northeastern Minnesota. Federal officials identified the entire Lake Superior iron-ore mining district as one of several underdeveloped or depressed regions around the nation, with President Kennedy ensuring that it was included in the map of blighted regions. The Iron Range was described as a region where the economy was "sagging badly" and "under-developed." It was lumped together with other blighted regions such as Appalachia and Pennsylvania's coal-mining regions. The IRRRB soon partnered with the ARA as the logical state agency to dole out federal aid.[28]

The ARA dramatically changed the IRRRB's role. Because federal funds were available, the state agency's major task was to attract federal grants for the Iron Range. As the IRRRB noted in 1966, "the most significant contribution of the [IRRRB] . . . to the State of Minnesota during the past biennium was the participation in federal programs which resulted in a total of approximately $510,000 in participating federal funds." Indeed, the lake states region of the ARA—an area including Illinois, Indiana, Michigan, Minnesota, Ohio, and Wisconsin—received the second-highest number of ARA projects and the third-highest amount of federal investment. Minnesota was ninth overall among states receiving ARA aid, ultimately receiving 2.7 percent of the ARA's spending.[29]

The ARA did not last long. The program was disbanded in 1965 when Congress allowed the original legislation to run out. The ARA's activities were subsumed by Lyndon Johnson's wide-ranging War on Poverty and the programs of the emerging Great Society. Many of the ARA's programs were folded into a new federal agency, the Economic Development Administration (EDA). For the Iron Range and the IRRRB, however, the brief flourishing of the ARA and the subsequent federal retreat from regional economic development were a path not taken in the nation's response to industrial decline. For a brief moment, the nation followed the route taken by the IRRRB decades earlier by committing to maintain a viable industrial economy in depressed areas such as the Iron Range. It is impossible to know how such a policy might have turned out had it been pursued over a longer period of time, but it nonetheless illustrates that the state and local focus of postwar economic development was not the only available option.[30]

In the second half of the 1960s, especially following passage of the 1964 taconite amendment and the subsequent expansion in taconite production, the IRRRB was challenged to formulate economic development plans in the midst of an economic boom. On the one hand, the taconite boom fulfilled the promise of economic development rhetoric and alleviated the unemployment crisis on the Iron Range. By creating a new taconite industry through the combined work of technology and public policy, economic development policy appeared to have proved its value on the Iron Range. Declining unemployment also turned down the pressure on development officials to provide immediate fixes for the regional economy and allowed them to focus instead on long-term development and diversification efforts.

On the other hand, it soon became clear that taconite would not be a panacea for the Iron Range's economy. Although building taconite plants employed a huge number of construction workers, the plants were highly automated once operational. They needed fewer workers than the natural ore mines and those workers needed different skills, meaning that unemployed natural ore miners could not count on moving into the new taconite plants without additional education or training. Thus, the new taconite mining economy obscured more deeply entrenched employment problems that continued well into the late 1960s, including an oversupply of workers unskilled to work in the taconite plants.

The IRRRB argued for continued economic development amid the taconite boom, but faced state officials and a public that did not understand the complicated structural unemployment now plaguing the Iron Range. Calls for continued development faced an uphill battle against legislators and a public that believed the Iron Range's economic problems had been solved by taconite. The IRRRB nonetheless pressed its case for ongoing economic development. In 1964, the IRRRB argued that the Iron Range needed vigorous rehabilitation efforts in the 1960s just as it had during the Great Depression. "Unemployment will still be a problem when the plants are completed and the taconite operations are in full swing," an agency report noted. "There will still be need for a rehabilitation program for area residents, displaced miners, [and those] considered too old and unqualified for the skilled and semi-skilled labor required by these plants." The IRRRB's point was echoed in the national press, with the *New York Times* reporting that although taconite construction was booming in 1965, state officials were worried that the construction boom was only a temporary uptick in employment. These fears were confirmed by the end of the decade. In a 1970 report, the IRRRB's commissioner noted that taconite had not resolved the underlying economic problems of the Iron Range: "The taconite industry has saved Northeastern Minnesota from an economic depression, but the area is still plagued by inadequate employment, job migration, and a lack of industrial diversity." The 1960s made it clear that taconite alone would not save the Iron Range's economy.[31]

New Priorities in the 1970s

With taconite production alleviating the worst of the region's unemployment problems during the 1970s, the IRRRB moved toward long-range planning rather than immediate relief efforts. Reporters visiting the Iron Range during the taconite construction boom described an industrial mining region that had averted economic disaster and was now enjoying prosperity. Commenting on taconite plant construction, a Hibbing official said in 1975, "I believe we have the biggest construction boom in the entire country." Boosters optimistically predicted population growth to accompany the construction boom, with some planning for an increase of twelve thousand new residents on the Iron Range. Reality proved more complex. The taconite boom certainly buoyed the spirits of Iron Range residents, but unemployment remained high throughout the 1970s. St. Louis County suffered from double-digit unemployment levels and was classified as "economically depressed" by Minnesota's Department of Economic Development. Population change on the Iron Range was also less spectacular than initial projections indicated. St. Louis County as a whole grew by only 1,536 residents during the decade, while population change was mixed in Iron Range towns. The two largest cities on the Iron Range, Hibbing and Virginia, went in different directions. Hibbing's population grew by almost 30 percent during the 1970s, while Virginia lost approximately five hundred residents.[32]

Free from the pressure to provide immediate relief for the unemployed and the poor, the IRRRB could focus instead on several long-range development plans such as environmental cleanup and tourism. As described in chapter 3, the environmental movement brought a new, if controversial, acknowledgment of mining's devastating effects on the landscape. Beginning in the 1970s, the IRRRB joined state-level efforts to clean up the worst environmental consequences of open-pit mining and to begin reclamation efforts on abandoned mine pits. In 1977, Minnesota's legislature gave the IRRRB control over the Taconite Environmental Protection Fund, a pool of tax receipts that was intended to fund cleanup, restoration, and rehabilitation of those areas of the Iron Range that had suffered the worst environmental damage from iron-ore mining. The IRRRB added a mineland reclamation division to manage its new environmental mission.[33]

In addition to major mine reclamation projects, the IRRRB undertook smaller efforts at beautification. Among the most popular of several such projects was a program to dispose of junked vehicles and destroy dilapidated buildings. The abandoned vehicle program and the building demolition program were intended "to remove unsightly and dilapidated buildings" and vehicles from the Iron Range landscape. As of 2002, the building demolition program had destroyed 6,550 old buildings across the Iron Range. These programs were successful, in part, because they offered free waste disposal for Iron Range residents hoping to get rid of old cars. Yet these programs revealed how decades of economic decline in a rural industrial region had left visible traces in the junk scattered across the landscape. Through programs such as the abandoned vehicle and building program, the Iron Range was trying to maintain the neat and orderly appearance typical of more prosperous regions. These programs were only partly successful. They did eliminate a great deal of the Iron Range's visible blight and have largely prevented the noticeable physical decay seen in Rust Belt cities such as Detroit or Cleveland. Yet a great deal of junk, especially old cars, remained on the Iron Range. "Junk defines the Iron Range," commented local journalist Aaron Brown. "Iron Rangers are a proud, noble people . . . who leave things in our yards."[34]

Developing the Iron Range's tourism economy was another major focus of the IRRRB during the 1970s. Officials had been discussing tourism promotion plans since the 1930s, but changes to the IRRRB's statutory authority and new sources of funding from taconite made tourism a priority for the IRRRB during this decade. Prior to the 1970s, a series of legal decisions limited the scope of the projects to which the agency could legally contribute. The IRRRB's original mandate emphasized promoting industry, and Minnesota's attorney general ruled that tourism did not count as an industry under the agency's legal mandate. In the early 1970s, however, the state changed its position and ruled that "tourism has a definitive economic impact upon the area, it is to be considered as a resource and, therefore, is in keeping with the purpose and goals of the department." With this decision, the IRRRB was now free to devote time and money to tourism projects on the Iron Range. By 1978, the agency reported that it was "deeply involved in the development, expansion and promotion of tourism as a viable industry to the economy of the Iron Ranges of Minnesota." Efforts to promote tourism on the Iron Range (described in greater detail in chapter 5) would remain a major focus of the agency throughout the last decades of the twentieth century.[35]

To better understand the various facets of economic development efforts during this period, the town of Ely, Minnesota, offers a useful example. Located on the rapidly declining Vermilion Range north of the Mesabi, Ely was a microcosm for both the audacity and the limits of the changes proposed by local economic development projects during the 1960s and 1970s. As Ely faced the end of its mining economy, economic development officials pursued different projects aimed first at attracting new industry to the town and, later, once those efforts had failed, projects to redesign the town's culture to meet the needs of a new service and tourist economy. While towns on the Mesabi Range were spared the worst of the declining natural ore mining industry resulting from the development of taconite mining, villages on the Vermilion and Cuyuna ranges were not so lucky. These areas did not have reserves of taconite ore and they were forced to confront the end of the mining industry during the 1950s and 1960s. The last mine on the Vermilion Range closed in 1967. Civic leaders in Ely first turned to the IRRRB for assistance in the 1950s as part of a plan to attract new industries. In 1959, the IRRRB prepared a prospectus intended to attract relocating businesses to Ely. The prospectus hinted at the town's desperation, however, claiming, in bold capital letters, "ELY IS DEFINITELY AND VITALLY DESIROUS OF INDUSTRIAL DEVELOPMENT." Additionally, the prospectus hinted that Ely had a reserve of unemployed or underemployed workers willing to work for low wages, and pointed to a pool of five hundred available workers that included two hundred women "available for part-time work." By highlighting a pool of workers trapped in the remote town, including women looking for part-time work, the IRRRB and Ely admitted that the only industries likely to relocate were those hoping to take advantage of a town struggling with industrial decline. During this first stage of Ely's economic development plans, the goal was simply to attract new industries that might offer employment and replace the wages lost in the mines.[36]

When industrial recruitment efforts failed to attract substantial new employers, Ely and the IRRRB attempted more ambitious economic development projects aimed at re-creating Ely's workforce and culture to make them amenable to a new economy based on service and tourism. The wide-ranging Title I Project, undertaken in the 1960s by Ely's civic planners in conjunction with outside consultants as part of the Elementary and Secondary Education Act, marked just such an attempt. The Title I Project brought together Ely's employers, schools, and local officials in an effort to develop the city's "human capital" for a postmining economy. The ultimate goal of the Title I Project was to change the attitudes of Ely's residents, encouraging them to focus on a positive vision for a postindustrial future.

The most important aspect of the Title I Project was an effort to develop Ely's human resources. In the late 1960s, officials in the area's public schools and junior colleges developed a liaison project to coordinate "the efforts, resources, and facilities of the Eveleth Area Vocational Technical School, Vermilion State Junior College, and the University of Minnesota, Duluth with the agencies that are working to improve economic conditions in Northeastern Minnesota." Trying to determine which community institutions could take a leading role in transforming Ely after the mining companies left the area, civic leaders latched onto higher education as a tool for economic development. Local vocational and community colleges could provide "catalytic leadership" to help solve "economic, social, personal, and educational problems in Ely." Imagined as focal points for building a region's human capital, local colleges became key tools in economic development officials' program for moving Ely away from industrial labor.[37]

Part of this change involved sprucing up the town's physical appearance. A team of architects visiting with city leaders noted that the entire town would need a "face-lifting" to improve residents' spirits. This was consistent with the experience of other Iron Range towns where planning officials proposed even more drastic changes to the physical appearance and layout of towns as part of their redevelopment efforts. For example, a 1971 planning report urged the city of Gilbert—a typical Iron Range town centered on a main street of commercial buildings—to abandon its older style of urban grid layout in favor of "a new subdivision layout" that could offer residents "a better quality of life" and attract new residents. One report from Ely's Title I Project argued that "the [Title I] project illustrates how a . . . project of short duration can through community attitude change turn a total community 'about face' in its self-image and develop its capacity to solve its own problems . . . Today, Ely can be identified as a city with a vision and a positive future rather than a city with only an image of itself in the past." Although such reports frequently overstated the amount of "attitude change" among residents, the multifaceted attempt to develop Ely illustrates the many directions in which local economic development policy was moving in the postwar decades.[38]

Toward an Entrepreneurial Ethos in the 1980s and 1990s

Concerns that the IRRRB had abandoned its earlier mission of job creation surfaced during the 1970s. At the time, some critics claimed that the agency's emphasis on quality-of-life programs was a distraction from more pressing economic development issues. One speaker at an IRRRB meeting in 1979 argued that creating jobs should remain the cornerstone of the agency's development efforts: "Unless you provide the job opportunities," other environmental or social programs would be ineffective. In addition to critiques of the agency's priorities, the IRRRB received criticism from other branches of the state government. A legislative audit in 1980 found problems with the IRRRB's handling of grant money, and said that "IRRRB needs to make significant improvements in handling grants to adequately fulfill their financial management responsibilities." The audit described a pattern of mismanagement and sloppy record keeping at the IRRRB that fueled perceptions that the IRRRB was little more than institutionalized pork-barrel politics. For example, the IRRRB failed to provide the state's legislative auditor or the governor with a full accounting of how it spent its grant money in 1978 and 1979. The agency's fundamental direction and institutional practices were under fire by the early 1980s, setting the context for wholesale reorganization during the steel crisis of those years.[39]

When the U.S. steel industry collapsed in the early 1980s and unemployment on the Iron Range rose well into double digits (discussed in more detail in chapter 5), the IRRRB abandoned many of its long-range goals to focus on providing immediate economic relief. Returning to its Depression-era roots, the agency financed several public works projects intended to put unemployed miners back to work at the federal minimum wage. Iron Range residents could earn money while working on municipal projects for a maximum of two weeks. The agency's planning priorities in the first half of the 1980s were made under "emergency conditions" as the Iron Range's high unemployment eventually led Governor Al Quie to call the state legislature into a special one-day session to address the problem.[40] The steel crisis of the early 1980s caused the IRRRB to set aside its earlier priorities of long-range economic development to instead develop projects that could immediately put the unemployed to work. The guiding principle of the agency during the crisis was a need for "action, not reflection."[41]

Among the most controversial of the relief initiatives undertaken during the 1980s was the public financing of a Hibbing wood products industrial park that attracted a chopsticks manufacturer. The chopsticks factory, as it came to be known, was initially heralded as evidence of how public financing could promote entrepreneurship in rural America during the age of globalization. But the plant soon ran into trouble and eventually became a tragicomic example of economic development boondoggles. When Hibbing built an industrial park to attract wood products businesses in the 1980s, Canadian entrepreneur Ian Ward jumped at the chance to use public financing for his scheme to build a chopsticks factory under the name of Lakewood Industries. Ward's plan to produce chopsticks for export to Japan was received with amusement by the national press. The Associated Press titled one story about the plant "High-Tech Chopsticks Factory Hopes to Make a Fortune, Cookie." Nonetheless, the opening of a manufacturing industry in Hibbing, where unemployment was rampant as a result of the steel downturn, led to a rush of applications from workers desperate for a job. More than three thousand hopeful Iron Range residents submitted applications for one of the first thirty-two spots in the factory. Even after the initial rush, several dozen people called the factory each day asking about work. The factory did not offer high wages, but it tapped into a pool of employees hoping to stay in the area and desperate for work. The average pay of $5.50 to $7.00 per hour was considered low in comparison to mining jobs.[42]

The plant was possible because of generous public funding from the IRRRB. Agency officials were initially hesitant to fund such an unusual venture. The IRRRB's development director described the chopsticks factory as "a little far-fetched, especially for the Midwest." The plant was touted by Governor Rudy Perpich, who claimed it was the beginning of a bright new era of manufacturing on the Iron Range, and the IRRRB eventually supported the project with substantial funding. Approximately 30 percent of the factory's initial start-up funding was provided by the IRRRB, a total of $3.4 million. Lakewood Industries received another million dollars in public money from other local sources. The city of Hibbing, desperate for industrial businesses, offered Lakewood ten free acres of land, almost a quarter-million dollars in grant money, and another half-million dollars in loans and tax-increment financing. The plant soon ran into trouble, however, when its technology failed to produce wooden chopsticks reliably. In January 1989 the plant closed down, putting sixty-five employees out of work. The plant was hugely in debt, losing all of the IRRRB's initial investment and several million additional dollars. After trying to restart in the winter of 1989, the chopsticks factory closed for the last time in July 1989 with several million dollars of debt, thousands in unpaid taxes, and dozens left unemployed.[43]

The Lakewood Industries debacle, along with similar industrial development schemes paid for with public money, soon became the focus of criticism. Local residents criticized the IRRRB and municipal agencies for trying to lure industry to the region with publicly financed industrial buildings, arguing that such efforts were likely to attract only low-wage employers. One Iron Range resident complained to the IRRRB that industrial parks brought exploitative employers to the Iron Range, telling the agency that such companies "paid scab wages in a Range town [and] caused more trouble . . . because of their attitude towards labor." If the IRRRB was truly committed to luring businesses to the area to alleviate economic suffering, the resident said, it should create "guidelines so that we don't get companies in there that exploit the people, [and] that don't pay a wage that's in line with the area."[44] Another critique came from outside observers in the field of planning analysis who argued that throwing public money at businesses likely to expand in the region meant that the IRRRB might very well be supporting enterprises that would have come to the Iron Range without any public money. Although the IRRRB funded large-scale development of a wood products industry in the region, for example, it was widely known that the market conditions for wood products were favorable and the Iron Range was a likely place for corporate relocations. Thus, "firms that have received subsidies . . . would have located to the region—or existing firms would have expanded—in response to information about good market and resource conditions, regardless of public financial assistance." In other words, the IRRRB was spending public money to support private businesses that likely would have relocated without any assistance but were happy to take the agency's funds. Controversial projects such as these ultimately led to a larger shift in the IRRRB's orientation toward economic development projects.[45]

Many observers felt that the steel crisis and high unemployment of the 1980s signaled a fundamental turning point for the Iron Range and the IRRRB. It was no longer possible, these observers said, for the Iron Range to count on the iron-ore mining industry to provide long-term economic growth and employment for the region. Although the IRRRB was founded amid similar worries during the middle of the century, the taconite boom of the 1960s and 1970s had lulled many Iron Range residents into a sense of economic security about iron-ore mining and the steel industry.

From the IRRRB's perspective, the slump of the 1980s brought renewed attention to economic development beyond iron-ore mining. For many years the taconite industry had alleviated the region's most pressing concerns about unemployment. But the crisis during the early 1980s reminded everyone that the Iron Range was still largely dependent on the mining industry for its economic health. "It has become increasingly apparent," the IRRRB's commissioner wrote in 1984,

> that dependence upon a single industry . . . is neither healthy
> nor economically viable. For while taconite is still important—
> and will continue to be important—it will never again be the
> single industry or activity which will support the Iron Range.
> It cannot and will not supply the dollars we need for education,
> for municipalities, for social programs. Because of this, the
> IRRRB is addressing the whole area of economic development
> and concentrating on a multi-faceted approach to the region's
> problems.[46]

Articles in the national press mirrored the commissioner's sentiments, as when a 1985 *New York Times* article suggested that the Iron Range's entire future was "iffy" given the collapse of the American steel industry and the rise in cheap imported iron ore from South America. "Local officials are struggling to revive the region," the article noted, "pursuing a policy of diversification forced upon them by their grudging recognition that the area's chief industry—its raison d'être—is in a seemingly inexorable decline." As the IRRRB planned for economic development policies in the years after the steel crisis, it was guided by the idea that it could no longer rely only on the iron mining industry as the primary driver of regional economic growth.[47]

A key document guiding the transition of the IRRRB's mission in this era was a comprehensive report of the IRRRB's activities by the consulting firm of Arthur D. Little, Inc. This long report pulled no punches about the future of iron-ore mining and steel-related industry on the Iron Range. The report emphasized that the Iron Range was suffering from "a major structural shift in its economy with the decline of taconite production" and warned that the taconite slowdown "is not a temporary or short-term phenomenon" but marked a new era of decreased production that would continue indefinitely. Attempts to entice new industry to relocate to the region would also be complicated by increasing regional competition for heavy industry. The report noted that the Iron Range was one of many regions trying to develop new industry and diversify its economy as a result of the 1980s recession. It thus warned that the IRRRB should not plan to bring new industries to the area by "smokestack chasing." Because the agency depended on taconite taxes to fund its operations, the report argued that decreasing taconite production meant that "the IRRRB's financial resources will be more limited in the future." The IRRRB was urged to limit the scope and variety of projects it funded. Overall, the report urged the IRRRB to "pursue an aggressive and accelerated reallocation of resources from their present broad based application to more targeted economic development efforts." The agency was urged to emphasize its role as a financier of new enterprise on the Iron Range, a role that it had the institutional capabilities to handle well.[48]

Pushed by economic crisis and outside consultants, the IRRRB turned in the late 1980s and 1990s toward a business development model that drew on a corporate ethos rather than its original state-centered, public-welfare model. In many ways, the agency came to resemble a corporation itself, emphasizing a new profit-minded approach to development and encouraging an entrepreneurial mind-set within the agency and the projects that it supported. The agency's biennial report to the state legislature even mimicked a corporate annual report rather than a government document. In part, the turn away from directly financing projects was motivated by high-profile failures among IRRRB-supported projects. Because of highly publicized incidents such as the chopsticks factory, some IRRRB officials grew concerned about lending money to Iron Range businesses for projects that had little chance of success. As one IRRRB member put it, "we have on so many occasions wound up with egg on our faces . . . we have funded people that have some project going on and nothing has ever come of it." Although it had always been important that the IRRRB put its money toward productive ventures, the agency gradually shifted to emphasize that government, by its very nature, could not allocate money as productively as private enterprise and should take a more hands-off approach in business development.[49]

One example of this attitude was a new financing program that removed the IRRRB and its pool of tax money from directly funding businesses and instead emphasized the agency's role as a backstop for private financing and mitigating business risk. Under a new bank participation loan program, the IRRRB would back one-half of a loan to an eligible business at a rate less than that offered by commercial banks. The overall goal of the program, as described in IRRRB promotional literature, was to offer "lower cost capital while simultaneously spreading the exposure among a greater number of lenders." In part, the IRRRB was able to support economic development loans in the late 1980s because it had a large new fund under its control. As part of changes in taconite tax laws made during the late 1970s, money from new taconite taxes was put into a special fund called the Northeastern Minnesota Economic Protection Trust Fund, better known as the 2002 Fund (later renamed the Douglas J. Johnson Economic Protection Trust Fund in honor of a state senator). Access to this money allowed the agency to increase its lending operations.[50]

The IRRRB's rhetoric at the turn of the twenty-first century contained a mounting hostility to direct agency involvement in the region's economic affairs and, at times, a gnawing sense that government itself was the problem. By the late twentieth century, many Iron Range residents imagined government as a lumbering, bloated, and inefficient entity. This vision of government sharply contrasted with earlier ideals of local government as a protector against the ravages of industrial capitalism and powerful corporate interests. The IRRRB took advantage of these new pro-business sentiments, marketing itself as a government agency that did not really act like the government. It was, according to IRRRB literature, a public–private partnership that worked, which described it with phrases such as "partnerships that work" and "partnerships for progress." The emphasis on the agency as a partner to business reflected the shift toward a belief that markets, not government, were the appropriate framework for economic development.

Entrepreneurship became a key term in the IRRRB's rhetoric during these years. By celebrating the entrepreneur as a potential savior for the beleaguered region, the IRRRB closely mirrored broad trends in American political and economic culture in the late twentieth century. As urban planning scholar Robert Beauregard notes, "economic development officials in the United States . . . have elevated successful investors, developers . . . and corporate executives to near cult status." Entrepreneurship also relieved the IRRRB of worrying about deep structural issues such as nagging unemployment or persistent rural poverty. If individual business geniuses offered salvation and renewed economies, then the role of economic development agencies was to entice and nurture entrepreneurs until one could revitalize the Iron Range.[51]

Indeed, one major critique of the IRRRB during this era was that it failed to promote an "entrepreneurial spirit" on the Iron Range. In a 1982 magazine article, a business journalist argued that the Iron Range's history of an "us-versus-them view of the world," conditioned by decades of isolation and labor strife, hampered entrepreneurialism among Iron Range residents. A lack of entrepreneurial spirit, he said, lay behind the region's failing economy: "There are at least 900 northeastern Minnesota businesses that sell products or services to the taconite industry, presumably, a good many of those products and services would be of use to mining and heavy-manufacturing operations around the world. What the arrowhead [region] seems to have lacked is entrepreneurs with the daring and desire to pursue those broader markets." The lack of "daring" entrepreneurs was not limited to industry. It also stifled the region's service businesses. "The Iron Range is famous for, and loudly proud of, its rich and diverse ethnic heritage," the journalist noted, "one might expect to find a good number of ethnic restaurants in the area, but they are very scarce. The region has apparently lacked entrepreneurs to launch such ventures." Entrepreneurship became a powerful tool for understanding economic problems during these years because it offered a simple explanation for what ailed the Iron Range and a panacea for curing a complex set of intractable economic problems.[52] By the end of the century, the IRRRB had turned away from its original public welfare model and instead presented itself as a flexible and business-minded institution that stood ready to promote entrepreneurship without intruding on business prerogatives.

Conclusion

The IRRRB's haphazard evolution as a regional economic development agency illustrates the uncertain path of government economic planning, especially at the state and local level, between the New Deal and the end of the twentieth century. The creation and growth of agencies such as the IRRRB during the postwar years suggests that comprehensive economic planning occurred at the state and local level in the immediate post–World War II era. As the federal government retreated from its activist role in economic planning in the aftermath of the New Deal, a variety of state and local agencies took up the challenge of coordinating an economy undergoing tremendous upheaval throughout the postwar years. In the case of Pennsylvania's anthracite coal-mining regions, for example, Thomas Dublin and Walter Licht argue that "for a decade after World War II . . . industrial development campaigns were strictly local undertakings." As historians such as Otis Graham, Thomas Sugrue, Thomas Dublin, and Walter Licht have explained, a plethora of development initiatives were born during this era, operating under a variety of titles, including economic development, rehabilitation, or urban renewal. From a national perspective, the IRRRB emerged as one of several economic development agencies created at the state and regional levels to combat industrial decline and ensure that the government had an active hand in shaping an economic future based on industrial production. The unique feature of the IRRRB, which perhaps explains its later success in achieving economic development, was the active involvement of state government, in contrast to other locales where economic development was exclusively local.[53]

The primary role of state and local agencies in economic development policy emphasizes the important, perhaps central, role that local government played even during the height of Cold War liberalism, a time typically associated with a rapidly expanding federal government. The key role of the IRRRB in coordinating economic development policies, including federal policies during the 1960s, is a reminder that federal liberalism during this era was often channeled through state and local agencies that proved crucial to shaping the direction and texture of liberalism. In short, understanding economic development nationally requires attention to state and local government agencies such as the IRRRB.

Unfortunately, as historian Thomas Sugrue argues, state and local governments too often "remain a terra incognita" for historians. Emphasis on national narratives has focused attention on federal agencies and presidential policies, often at the expense of state and local governments that often proved more consequential to the fate of those policies and the lived experience of government during the twentieth century. As Sugrue explains, by the middle of the twentieth century the federal government's power had expanded to reach into nearly all aspects of American life. "To an extent unimaginable in the nineteenth century," he notes, "ordinary Americans lived life in the shadow of the state." Yet federal power typically arrived in the form of local officials, often driven by complex motivations.[54]

Relying on state and local government to handle economic development had lasting consequences on the contours of American industrial decline and the lives of people remaining in depressed regions such as the Iron Range. The emphasis on state and local governments in economic development planning cut off national options that might well have had far different results. In contrast to the American example of local responses to industrial decline, historian Steven High describes a nationalist response to deindustrialization in Canada's manufacturing heartland. Canadian workers facing downsizing and layoffs called on the national government to protect their jobs, and in many cases provoked a national outcry over Canadian deindustrialization. The tension between the local and the national also had consequences for the political culture in industrial regions, where many workers have oscillated wildly between liberal, and even radical, responses to their depressed communities and a jingoistic patriotism that celebrates the United States as a land of prosperity and opportunity.[55]

Urban planning scholar Robert Beauregard writes that "economic development seems like such an appropriate thing to do, regardless of one's political persuasion, that one cannot criticize without being seen as a nay-sayer and an opponent of progress." But, as this chapter suggests, economic development has often proceeded with surprisingly little critical or strategic concern. What type of future should policies try to develop? Of the many divergent interests in any given community, whose concerns should be prioritized when making development policies? Answering these kinds of questions might well make for a more democratic and useful future for economic development, on the Iron Range and elsewhere.[56]

FIVE

The Turn to Heritage

Conflicts over Mining's Memory

The Iron Range's struggle against decline played out in laboratories
and tax commission meetings in the decades after World War II.
But the Iron Range's post–World War II history was also a saga of
memories about the region's past and plans for the future. These
two visions, one looking backward and the other facing forward,
came together in debates over heritage tourism and industrial nostalgia on the postwar Iron Range.

163

There were many different efforts to promote industrial heritage on the Iron Range during the last decades of the twentieth century, including turning abandoned mines into historical sites, re-
creating old mining villages, and building a comprehensive mining
museum and interpretative center to make the Iron Range's history
both accessible and marketable. Deep cultural and political tensions ran beneath industrial heritage programs and other efforts to
promote tourism on the Iron Range. Industrial heritage programs

often pitted two groups against each other. On one side was a small but enthusiastic core of local government officials, regional planners, and owners of service-sector businesses who were convinced that the Iron Range needed to develop a diversified, postmining economy and saw industrial heritage and the related tourism industry as the best path to achieve this goal. On the other side was a loose group of disgruntled industrial miners, only some of whom still worked in the active mines, and community members who believed that the Iron Range was a productive industrial region that should remain a valuable producer of raw materials for the steel industry. These two groups proposed radically different futures for the Iron Range during the late twentieth century. Those promoting industrial heritage saw a region moving past its industrial history and building a new culture and economy based on service-sector businesses, while those hoping to remain an industrial mining region wanted to expand existing mines and devote the necessary public resources to ensure that the Iron Range would remain a productive iron-ore mining region into the future. Similar conflicts over divergent futures have played out across industrial America from the 1970s to the present day.

Industrial heritage's role in conflicts over the ultimate meaning of the Iron Range raises difficult questions about the political consequences of history in deindustrializing regions. The Iron Range's experience in developing industrial heritage programs suggests that using history as an economic development tool can depoliticize difficult conflicts over the meaning of the past and the direction of the future. By promoting the Iron Range's heritage of industrial mining, the concerns and dreams of current miners were removed to an apolitical realm of nostalgia where the industrial past was celebrated but also cut out of debates about the directions of a postindustrial future. In other words, industrial heritage on the Iron Range celebrated industry but removed it to a nostalgic past where it had little to say about the present or future. On the Iron Range and elsewhere, efforts to promote industrial heritage as a tool for reviving declining regions highlight how planners and historians have participated in removing industry from narratives about modernity and vibrant national futures. As cultural critic Carlo Rotella observes, redevelopment plans based on celebrating the industrial past contain an "untroubled fatalism" about an industrial future. Historian Mike Wallace similarly notes that industrial history museums often relegate industry to a nostalgic past that has little to say to a postindustrial present or future.[1] If industry is already surrounded by the rich, rusted patina of heritage, how can it play any vital role in the future of individual workers, the Iron Range, or the nation?

Efforts to memorialize mining on the Iron Range were part of a multifaceted, transnational movement in the late twentieth century to revitalize declining industrial regions through tourism, heritage, public history, and culture-sector work. While postindustrial revitalization efforts took many forms, almost all shared an assumption that history, especially the physical remains of past industry, offered useful raw materials for marketing an area as a distinct place with a unique past—a heritage—worth visiting. History, especially historical structures, became a key element of postindustrial revitalization plans across late-twentieth-century America. Structures such as abandoned factories, Rotella writes, "become the basis of 'imageability,' a quality of combined depth and sharpness in both a place's physical appearance and its cultural reputation." Behind the importance of "imageability," which made industrial sites potent markers of heritage, was a concern with creating authentic and unique experiences for tourists in a seemingly inauthentic world of interchangeable places. The turn to industrial heritage as a building block for authentic tourist experiences was based on a belief that "visitors have to know how to get there, and they have to know they are somewhere once they do get there."[2]

Several examples, drawn from the diverse work of historians, anthropologists, and literary scholars, highlight the sweeping nature of these redevelopment efforts and help to locate the Iron Range's heritage plans in a broader context. Heritage development in Lowell, Massachusetts, offers a telling example of how public history could be a tool for urban revitalization and the contradictions inherent in such efforts. A hub for textile manufacturing in the nineteenth century but deindustrialized by the mid-twentieth century, the city of Lowell was a forerunner in what has come to be known as "culture-led revitalization" in postindustrial regions. This type of revitalization is premised on promoting cultural spaces and events, such as heritage sites, unique shopping districts, and arts festivals. In Lowell these efforts centered on Lowell National Historical Park, a national park created expressly to celebrate Lowell's industrial heritage and, paradoxically, to promote its postindustrial future. Several trends contributed to the new emphasis on heritage as a development strategy. First, heritage projects were supported by changes within the history profession, especially in the United States, during the last third of the twentieth century that gave rise to a new class of professional public historians concerned with community engagement and conveying history to nonacademic audiences. Public history projects were also bolstered by funding made available for heritage projects across the country through the National Historic Preservation Act of 1966, and related legislation that promoted history and national heritage. Various state initiatives also supported and funded heritage projects throughout the 1970s and 1980s. In 1980, for example, the Minnesota Historical Society urged local agencies to preserve Minnesota's architectural heritage. These initiatives, ranging from the federal to the state level, intensified the movement for architectural preservation and heritage promotion at the end of the twentieth century.[3]

Many of these public history and heritage projects were top-down affairs, coordinated by small cadres of public history and city planning officials. As historian Cathy Stanton notes in the case of Lowell, "the newly dominant mode of place-making in much of the contemporary world is a highly professionalized and rationalized one. It . . . is more likely to be controlled from planning and design departments than from neighborhoods."[4] On the postwar Iron Range as well, efforts to promote historical tourism were controlled by a small number of officials who relied on a body of professional knowledge and rarely considered the wishes and alternative plans of those who prioritized mining over tourism, leading to tension throughout the postwar decades.

Cities such as Lowell that had long ago lost their industrial base did not have to confront the wishes of active industrial workers when planning heritage projects. In contrast, regions trying to promote postindustrial development in the midst of active industry, as was the case on the Iron Range, faced additional complications. Anthropologist Kathryn Dudley describes a project in Kenosha, Wisconsin, where city officials and urban planners tried to market the city as a bedroom community for commuters to Chicago and Milwaukee rather than a blue-collar manufacturing town. As Dudley notes, many of Kenosha's white-collar residents, including civic leaders, teachers, and business professionals, were deeply hostile toward the "blue-collar mentality" they believed permeated the local automobile factory and the union. Recognizing that union autoworkers received high wages—often as high as or higher than many white-collar professionals—for what many professionals considered low-skill work, some in the city's elite felt that blue-collar workers were purposefully holding back the city's development for fear of losing their central place in the city's cultural life. In the end, Kenosha's civic leaders used a trope of economic progress and a historical romanticization for the waning industrial era to push development schemes that moved the city away from auto manufacturing and labor unions. What is especially significant in Kenosha's example is the way in which history was invoked to overcome the objections of industrial workers to postindustrial development plans. By promising to celebrate Kenosha's industrial history, city planners pacified industrial workers while moving ahead with their plans to displace them from their central position in the city's life.[5]

Carlo Rotella offers another perspective on industrial heritage projects in his account of a failed effort to promote historical tourism in the blue-collar city of Brockton, Massachusetts, in the 1990s. Brockton, a former shoe-manufacturing city, hired an internationally famous landscape artist to design a master plan for the city. The artist's ambitious proposal centered on boxer Rocky Marciano, a famed native son of Brockton, and involved restoring the Marciano family home and creating historical tours based on his running routes that honeycombed through Brockton. The core idea was to create "a green cultural center that both celebrated and laid to rest [Brockton's] industrial past" out of "the atmospheric ruins of the mill city that once upon a time produced shoes and Rocky Marciano." Brockton's redevelopment was quickly complicated by tension between industrial workers and those supporting postindustrial redevelopment, as well as fissures between proponents of redevelopment who held competing visions. On one side were promoters of heritage who argued that celebrating the city's industrial heritage of working-class sons who made good, such as Rocky Marciano, "could provide not only a hook to attract interest . . . but also a guiding impulse of cultural continuity in a time of disorienting change." These promoters were stymied, however, by strong resistance from other planners who were skeptical of drawing on the industrial past altogether. From their perspective, "an aggressively orthodox insistence on the industrial past and its legacies contributes to Brockton's failure to come to terms with a new urban era." As Rotella explains, opponents of industrial heritage believed the city had to "shed or denature its traditionally forbidding aura of working-class toughness, especially because over the last couple of generations that aura has modulated into a dead-end reputation typical of depressed Rust Belt cities." Interestingly, these planners also proposed a plan based on the city's past. Rather than Marciano and shoe manufacturing, this camp hoped to emphasize the city's connection to Thomas Edison—who had a laboratory in Brockton—as part of a long history of "high-tech" innovation in the city. As these examples illustrate, attempts to redevelop declining industrial areas through heritage were widespread in the late twentieth century. In many instances, though, invoking heritage as both the tool for and the outcome of development created deep tensions about the past, present, and future of industry within local and national culture.[6]

Efforts to promote industrial heritage on the Iron Range connected with other broad movements in the late twentieth century, especially a small but growing effort to preserve the artifacts and material culture of early industry. Such efforts have taken many forms, but they typically emphasize preserving the relics of industrial labor and technology, including abandoned factories, machinery, or mine pits. Although there were scattered industrial preservation efforts in the mid-twentieth century, professional efforts to preserve industrial sites began with the creation of the Society for Industrial Archeology in 1971. Preservation of industrial buildings and sites soon connected with planners interested in using such sites for broader regional or urban redevelopment plans. Trying to create authentic places for visitors, regional and urban planners began developing increasingly complex and dynamic industrial heritage projects, such as the numerous industrial heritage trails that have sprung up in North America and Europe.[7]

Industrial heritage on the Iron Range emphasized preserving abandoned open-pit mines and the enormous mining machinery for public display. Landscape critics Peter Goin and C. Elizabeth Raymond have labeled such efforts to put mining machinery on display as "the trope of the huge tire." As practiced on the Iron Range, they note, "industrial tourism . . . enshrines the massive steam shovels and trucks that were the tools of transforming the landscape and builds viewing platforms for the abandoned pits." For many casual visitors to industrial heritage sites on the Iron Range, however, the ultimate meaning of the mines remains vague. The Iron Range is presented "alternately as a technological wonderland or a recovering natural area." While industrial heritage offered tantalizing possibilities to planners hoping to transform the Iron Range into a postmining tourist destination, the reality of industrial heritage sites in the area, and elsewhere, offers visitors a conflicted and vague narrative about the history and legacy of iron-ore mining.[8]

Finally, heritage tourism and public history on the Iron Range demonstrate how deindustrialization and the transition away from industrial labor was experienced as a cultural shift as well as an economic transition. Many industrial workers experienced deindustrialization as a feeling of being removed to the realm of history and nostalgia—members of a group whose time had passed and now would slowly rust into obscurity. What made this transition particularly wrenching for many industrial workers was how far it moved them from the experiences of the early and middle decades of the twentieth century, when many firmly believed that their lives and their labor were an indispensable part of national modernity.

Postindustrial developments, including heritage and tourism projects, have often resulted in the cultural displacement of industrial workers to the margins of American culture and society. This was clear in industrial workers' attitudes toward postindustrial development plans in Kenosha, as documented by Kathryn Dudley. Reflecting on these workers' responses to the new, postindustrial Kenosha, Dudley observes: "Many of the cultural symbols, beliefs, and values that once fortified a sense of moral order in our capitalist economy have been cast into doubt." Ultimately, she says, manual workers have become a new type of "American primitive" who stand outside the progress of modernity. Historian Steven High and photographer David Lewis similarly describe deindustrialization as a cultural process that involves "the displacement of industry and industrial workers to the cultural periphery." As a broader consequence of deindustrialization, "industrialism has lost its cultural centrality in North America. Industrial workers who once inhabited a central place have been displaced to the periphery: they have become outsiders looking in."[9] Themes of displacement and "outsiders looking in" run throughout the history of the Iron Range in the late twentieth century. Indeed, a personal and communal struggle against marginalization was a defining theme of the region's postwar history.

Tourism and the Service Economy on the Iron Range

Throughout the twentieth century, there were scattered calls to turn the Iron Range into a tourist destination. As early as the 1930s, visitors to the Iron Range noted the region's north woods beauty and imagined a postmining tourist economy for the region. A sociologist working on the Iron Range during the 1930s commented that tourism would be the logical response to the eventual decline of mining in the region. He presciently predicted that as mining declined, civic boosters would promote tourism by "advertising of the hinterlands as a summer vacation center of unusual attractions." The sociologist was pessimistic about tourism's long-term viability on the Iron Range: "while this seasonal trade [tourism] stimulates business, it cannot assure permanence to the community." Writing half a century before widespread efforts to promote Iron Range tourism, he anticipated several of the major tensions that would bedevil Iron Range tourism plans. Although tourism offered economic possibilities for a remote region, it was ultimately a seasonal trade that relied on wealthy visitors from outside the region.[10]

By the middle decades of the twentieth century, a small but important tourist industry had developed in northeastern Minnesota. A travel reporter for the *Saturday Evening Post* visited the region in 1961 and was surprised to learn that tourism was Minnesota's third-largest industry, fueled largely by a lucrative recreational fishing trade. The reporter noted with some amusement that northern lodges were attempting to extend the tourism season into the winter by promoting winter sports such as cross-country skiing and snowmobiling, but he found little local interest in winter activities beyond ice fishing. Tourism was indeed playing a larger part in the Iron Range's economy by the 1950s and 1960s. A 1961 survey found that 5,810 employees worked in resorts throughout St. Louis and Itasca counties, a number that was increasing as mining employment was falling. A 1965 advertisement by the Minnesota Power and Light Company promised nothing less than "a new era of prosperity" for northeastern Minnesota based in part on "a new revitalized tourist and resort business" and "expanding cultural and recreational facilities." Many observers and planning officials saw tourism as a possible savior for the iron-ore mining areas once mining was finished.[11]

Throughout the 1950s and 1960s, local and state government officials supported plans to create a tourist economy on the Iron Range.

As historian Aaron Shapiro notes, tourism in the upper Great Lakes relied heavily on local and state government support: "Government at all levels facilitated tourist development through promotion, conservation, and construction efforts—becoming a vital player and stakeholder in the tourist industry."[12] Among the most important initiatives that government agencies undertook to support Iron Range tourism were numerous surveys that described in great detail the potential market for tourism and the amount of money available if the region could lure tourists. The IRRRB used the results from one such survey in the late 1950s to argue for further government action to build an infrastructure for tourist development in the region. Specifically, the IRRRB recommended forming "private and public associations" to "strengthen programs to encourage the improvement and rebuilding of resorts, hotels, motels, and camping and trailer courts, and their recreational facilities in the area." The survey report also called for tourism management courses at local high schools and community colleges and programs to lend money to tourist businesses. Later surveys went into far greater detail about the type of vacationer who might come to the Iron Range if appropriate facilities were built. One survey described a "typical vacationing party" that might travel to northeastern Minnesota as being "very attractive—young, well-educated, upper income, and in family parties." Readers of the report were invited to imagine these attractive parties visiting tourist projects supported by local and state officials.[13]

One recommendation that became reality was resort management courses in local community colleges and high schools. An expanding tourist economy needed workers with far different skills than those encouraged by the mining industry, especially customer service abilities. By creating courses in resort and tourist training, Iron Range schools hoped to train a new generation of workers for service-sector jobs. Courses began in the early 1960s when Iron Range junior colleges created a Resort Management Institute with the goal of training "tourist-oriented business operators." The colleges also launched ongoing certification courses for workers in the tourism sector. In Iron Range high schools, the economic logic of the tourism industry was made explicit for the students. A 1968 report for Iron Range high-school students, for example, began with pleasant phrases about how the area's natural resources could be harnessed for tourism. "The natural beauty of the area, its fresh air, recreational opportunities, historic monuments and other tangible and intangible things which attract people's curiosity and desires can and do become a source of income for the area by being 'sold' to tourists coming to enjoy them." The report then bluntly stated tourism's importance as a replacement for the declining mining industry. "No single large source of income has been found to replace or ease economic conditions when the iron ore production declines," students were informed, but "an expansion of tourism would help to diversify this area's source of income." Although planners often touted tourism's benefits, the industry was also pushed on young Iron Range residents as one of their only options to stay in the region. Young people on the Iron Range were told to embrace a tourist economy or face the prospect of unemployment, decline, and dispersal from the region.[14]

An example from 1965 demonstrates how developing tourism sites in the mining region proved far more difficult than initially presumed. While tourism in the abstract was a boon to the Iron Range's struggling economy, the messy realities of Iron Range geography, economics, and culture made new tourism projects complicated and expensive. One major tourism project planned for the Iron Range in the 1960s was the development of a recreation and resort area on a lake outside Hibbing. Suggested in a 1965 study, planners recommended developing six thousand acres around Carey Lake as a "medium density recreation development." By turning an unused area into a planned landscape with campgrounds, a supper club, a lodge, a golf course, hiking trails, and wildlife habitat, the Carey Lake project would hopefully meet "the present and future demands generated by tourist-vacation traffic." Planners also admitted that the development "would certainly help improve the economic base of that portion of the Iron Range region presently lacking adequate facilities for the tourist industry." Despite optimistic promises for the Carey Lake development, the feasibility study contained site information that was critical of the project's overall prospects. Reviewing the site and competing recreation areas, a consultant suggested that the development was not likely to be "a financial success." The site was problematic, as the proposed area was made up mainly of "vast, low, wet, peat swamps" that hardly met a tourist's vision of a sublime north woods landscape. Because of the swamps, "only a small percentage of the total site could actually be considered for intensive development. The remaining portions of the site are generally in a swampy condition at least parts of the year." Carey Lake itself was not particularly attractive. The consultant's report noted that the lake "does not have the attractiveness of most north country lakes" and was unlikely to draw tourists away from other, more scenic lakes in northeastern Minnesota. Although promoters were eager to build an infrastructure for tourism on the Iron Range, many sites were simply not attractive for tourism. The Iron Range existed because of iron-ore deposits, not its natural beauty.[15]

Given the difficulties in building tourist sites that would appeal to vacationers looking for natural beauty, state and local planners explored the possibility of using the region's industrial mining history to promote tourism by the late 1960s. In 1969, Minnesota created the Iron Range Trail across northeastern Minnesota. The trail was intended to connect Minnesota's three iron-ore mining regions and "related points on the North Shore of Lake Superior." It was hoped that hikers, bicyclists, and other sightseers would use the trail to explore northeastern Minnesota. As described in a promotional pamphlet, the trail would "provide the public with an opportunity to view and understand iron mining in the state" by marking "scenic and recreational areas as well as historic sites and local tourist centers where visitors may obtain more details on the area's attractions." Although the trail was controlled by the Minnesota Department of Natural Resources, the IRRRB immediately saw the tourism potential of the trail and made it the center of a growing tourism development plan.[16]

Although the Iron Range landscape spoiled plans at Carey Lake, other tourism projects were complicated by residents' ambivalence and even hostility to tourism in general. For miners who thought of the Iron Range as a hardworking region that provided critical raw materials for a vital industry, plans to transform the area into a haven for wealthy outsiders were diametrically opposed to how they understood their home. There was a complicated irony in this stance, of course, because many residents were acutely aware of their region's dependence on the outside capital of the mining companies. But many believed that through labor unions and hard-fought local politics, the Iron Range had successfully won control against outsiders. Appealing to a new group of wealthy out-of-town tourists upset this delicate balance.

Hostility to tourism rarely appeared in the formal reports of planners and development officials, yet close attention to the voices of Iron Range residents reveals a deep ambivalence about promoting tourism during this era. One planner was surprised to discover that "many of the individuals on the Range . . . expressed skepticism about tourism development." The planner argued that "the lack of strong, vocal community support for tourism on the Iron Range may well be the most difficult of all problems to overcome. Community support and patronage cannot be mandated. They will occur when, and only when, residents come to believe that support for tourism is in their own best interests."[17]

The hostility that many residents felt toward tourism was based on deeper concerns about working in low-wage service-sector jobs and cultural beliefs that the Iron Range should remain a productive landscape dedicated to resource extraction. Although tourism's boosters imagined resorts as viable options for a postmining economy on the Iron Range, many residents knew that work in the tourism sector paid substantially less than unionized mining jobs and offered even less job security. By the 1960s, many residents had personal experience working in the tourism industry or related service-sector work. The ongoing decline of mining employment meant that more and more Iron Range residents could expect to work in nonmining jobs during their working life. Of course, women on the Iron Range had a longer and more substantial history of nonmining employment given their discrimination from mine work until the 1970s.[18] But the turn toward service-sector employment was widespread by the 1960s. In 1966, for example, such work became the predominant employment category on the Iron Range, meaning that mining was no longer the region's largest source of employment. Iron Range residents were well aware that service-sector jobs like those in the tourism industry paid, on average, only one-third as much as mining work. Additionally, residents were worried that the tourism industry could not grow quickly enough to make up for the loss of mining jobs. A 1964 newspaper article described attempts to build ski resorts in the Lake Superior mining area and emphasized its natural beauty, but warned that "the tourist industry is not growing fast enough to offset the loss of jobs and income caused by the decline of mining." The tourism industry also contributed to economic inequality. Many miners believed, whether accurately or not, that tourism's promoters were working with resort owners who stood to profit disproportionately from a new tourism industry on the Iron Range. One laid-off miner described such a worry when explaining why he was hesitant to develop the Iron Range as a tourist destination. He claimed to have nothing against tourism itself, but worried that tourism would do nothing to alleviate economic decline on the Iron Range and would lead to further economic inequality in the area because tourism facility owners would benefit disproportionately. Although proponents of tourism saw it as a panacea for the ailing Iron Range, many area residents had legitimate concerns about the economics of turning the Iron Range into a tourism-based economy.[19]

In addition to these pocketbook concerns, turning the region into a tourist destination contradicted many Iron Range residents' understanding of the Iron Range as a productive, extractive landscape. To the eyes of many residents, the Iron Range's beauty lay not in the wooded landscape but instead emerged from the combined efforts of human labor and technology to dig iron ore and transform it into valuable iron and steel. This view of the landscape was difficult for many outsiders to understand. The Iron Range appeared to many outsiders as a tremendous scar, a fully artificial landscape where humans took the land and "flipped it like a pancake." Geographer Thomas Baerwald describes the Iron Range landscape as "long, reddish gray gashes [that] marked the sites where miners had torn iron ore from the earth." But many Iron Range residents saw this same landscape as visible evidence of many decades of productive labor. The pits reminded them of the generations of workers who had made this out-of-the-way place their own. Industrial miners on the Iron Range were not alone in their ambivalence or even hostility to reinterpretations of the industrial landscape for tourists' consumption. Steven High and David Lewis note that across North America working people have expressed deep uneasiness toward the movement to preserve historical buildings. This movement, they claim, "at its core, valued industrial buildings for their architectural beauty and for their potential reuse as post-industrial spaces," a mind-set that ultimately displaced industrial work to an outmoded past. "The factory-scape might be retained," they note, "but the jobs are gone, as were the workplace cultures on which industrial workers depended for status and solidarity." Driven by concerns over the low wages, employment insecurity, and conflicting understandings of the physical landscape, many Iron Range residents approached tourist promotion efforts with deep skepticism.[20]

Despite these worries, by the 1970s the IRRRB and planners in the area were committed to creating a tourist economy on the Iron Range. IRRRB officials were lured by the prospect of tourist dollars flowing in. Several major tourism and recreation plans were created in northern Minnesota during this period and Iron Range officials wanted to ensure that their region received part of the tourism trade. The construction of Voyageurs National Park in far northeast Minnesota was an especially significant catalyst for the IRRRB's tourism promotion efforts. Voyageurs National Park was established in 1975, creating Minnesota's first national park and a major new tourist draw in northern Minnesota. IRRRB officials imagined tourists streaming to the new park and began thinking of ways to divert them. At a 1973 IRRRB meeting, an executive with the Minnesota Department of Economic Development explained how travelers to and from the park could contribute to the Iron Range's economy:

> We have a captive audience to build from and grab the tourism dollar. In excess of ninety per cent of [the] million people visiting Voyageurs Park are going to take Highway 53. That affords an excellent opportunity to funnel them off into the Range area and we feel it is vital that they do be funneled off for two reasons: Number one, to give those people a varied experience as far as a tourist experience . . . and also the fact that the Voyageurs Park is going to realize a rather short season and if we put all those people in that park in that short period of time we are going to experience tremendous problems in that area.

Officials with the IRRRB were less enthusiastic about recreational tourism in the park but nonetheless saw the tourists as a potential source of revenue for the Iron Range. As one IRRRB member in 1975 explained:

> All the talk of the Voyageurs National Park has been in the news and all the tourists that are going to be in the area. I think that between the mosquitoes in June and the snow in July up at Voyageurs Park that these people are going to be leaving that area . . . they are going to be leaving after four hours. And any way we can quote, unquote, delay them in northeast Minnesota before they come back to the bright lights of Minneapolis-St. Paul is to our benefit.

Driven primarily by dreams of tourist dollars, Iron Range officials turned to tourism as a way to diversify the region's economic base by the 1970s.[21]

A viewing stand overlooking the Missabe Mountain Mine in 1937. Early visitors to the Iron Range were guided to stand in awe at the size and complexity of open-pit iron-ore mining. Photograph by Theodore Salmi. Minnesota Historical Society Collection, HD3.112 p59.

The Iron Range Interpretative Center

But what kind of tourism industry should—or could—be created on the Iron Range? After several years of conflicted debate, the Iron Range finally settled on an answer: heritage tourism. Given Iron Range residents' deep ambivalence about tourism in general and the reality that the Iron Range's physical landscape had been drastically upended by open-pit mining, industrial history emerged as a likely building block for a tourist economy during the 1970s. Attempts to promote heritage tourism on the Iron Range reveal the possibilities and contradictions of using history to promote economic development in postindustrial areas such as the Iron Range. Heritage tourism on the Iron Range was, as one report described, "a unique attempt to build a tourism base—where no base previously existed—by concentrating public monies in the preservation and promotion of the region's rich natural, historic and cultural resources."[22]

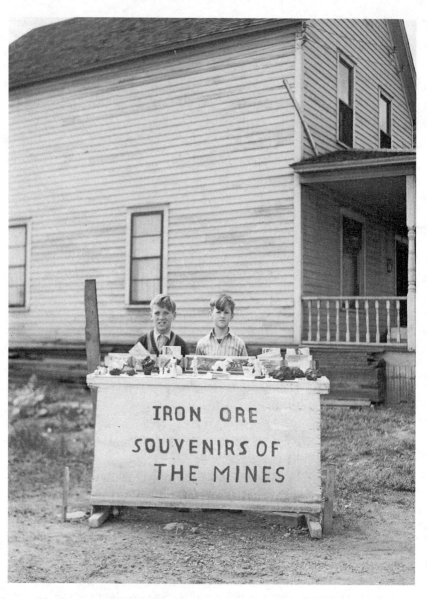

Children in Hibbing sell souvenirs from the iron-ore mines in 1941. Early tourism businesses on the Iron Range were small and focused on celebrating industrial mining. Photograph by John Vachon. Library of Congress, Prints and Photographs Division, FSA/OWI Collection, LC-USF34–063998-D.

Planners interested in developing a postindustrial tourist economy began exploring possibilities for heritage tourism on the Iron Range by the early 1970s. For instance, a 1972 report by an architectural and planning agency emphasized how industrial tours of the Iron Range could promote economic development: "Industrial processes are intrinsically interesting to nearly everyone." Noting the scattered tours then available in iron-ore mining pits—both operational and defunct—the report claimed, "all of these tours are important generators of tourist interest and deserve more widespread attention." It called for additional tours to interpret "the Iron Range story." Possible options included bus tours of historical and industrial sites or tours for "rock hounds" interested in geology. "Rail buffs" might be attracted to rail tours across the Iron Range. The long tradition of factory and mine tours may well have given planners the idea for broader development of tourist sites emphasizing industry and mining history.[23]

The Iron Range Interpretative Center (later known as Ironworld USA; the Ironworld Discovery Center; and the Minnesota Discovery Center) soon emerged as a hub for many of the Iron Range's heritage tourism efforts. The interpretative center—known as Ironworld to Iron Rangers—is described by one Iron Range journalist as "one of those strange places that you have to see to believe."[24] The Chisholm, Minnesota, facility has served over the years as a museum, a convention center, an archive, a concert facility, a food court, and a festival center. It has been asked to serve many different—and often contradicting—masters.

An interpretative center emerged as an idea among planners and state officials in the early 1970s. In initial plans sketched out at the IRRRB in 1972, the Iron Range Interpretative Center was conceived as a flexible, multipurpose site that "will tell the story of Minnesota's iron mining region. This will include the natural and human history of the area in addition to the mining aspects of this fascinating story." In part, the interpretative center grew out of the Iron Range Trail project. Officials felt that the trail, while useful, did not have definite starting or ending points and was not connected with the Iron Range towns it passed through. Thus, planners initially envisioned that the interpretative center "will provide the answer to the difficult question, 'Where does the Iron Range Trail start and end?' It will eliminate the confusion of the present system which does not provide direction to the visitor who is interested in the Iron Range Trail."[25] Travelers on the Iron Range Trail would hopefully use the interpretative center to coordinate their travel and connect the trail's many attractions with the Iron Range. Yet the insistence that the Iron Range Trail have a single, central point of information also reveals continuing uneasiness about tourism among Iron Range officials. Behind the center's initial plans was a fear that tourists on the trail would not recognize or acknowledge the region's mining history or, at the least, that the center was an effort to exert greater control over tourists' experiences.

Plans for the center quickly grew and it was soon envisioned not just as a site for tourist information, but as a multipurpose facility that would be a bustling hub of tourism and heritage travel on the Iron Range. In their 1972 report to the state, IRRRB planners described their expanded vision for the center:

> The proposed major Interpretative Center will be an imaginative, flexible and dramatic facility. Major emphasis will be placed upon participatory and self-guided activities. Flexibility is a key element. The facility will have to provide space and ways to do everything from handing out literature to exhibiting enormous pieces of mining equipment. There are four major categories of functions the Interpretative Center will provide: 1. Geological and natural history interpretation. 2. Historical and cultural interpretation. 3. Mining industry interpretation. 4. General visitor information service and relaxation facilities. It is important to point out that the proposed major Interpretation Center should not be regarded, merely, as a museum or collection of static displays. Among the methods that could be employed are models, audio visual presentations, photographic techniques and participatory activities as well as imaginative displays and exhibits.[26]

By preparing plans for a "dramatic" historical center that combined Iron Range history with tourist information, planners sought to simultaneously generate a tourist economy and ensure that any potential tourists celebrated the region's mining history.

Planners also imagined the interpretative center as a significant source of revenue for beleaguered Iron Range communities. Initial projections imagined it bustling year-round with tourists. Needing to sell the expensive facility to both the state legislature and Iron Range residents, planners at the IRRRB projected hundreds of thousands of visitors flowing through the center and out into the Iron Range towns where they would spend their money and support local businesses. Initially, planners projected that an Iron Range Interpretative Center would draw as many as 225,000 visitors to the region annually. This projection, which proved wildly optimistic, was based on the annual attendance figures from Chisholm's smaller Minnesota Museum of Mining, which drew approximately thirty thousand visitors per year in the late 1960s and early 1970s even though it was only open for three months a year. Attendance projections grew along with plans for the interpretative center. By 1974, projections anticipated a quarter million visitors per year, a workforce of twenty-one employees, and $12.9 million pumped into the local economy. Promises of visitors and the money they would bring were a key argument put forth by the IRRRB to justify the expensive center. By projecting thousands of visitors bringing dozens of jobs and millions of dollars, planners allowed both state politicians and Iron Range residents to imagine a busy center that worked like a valve: bringing a steady flow of outside money to the Iron Range, but also controlling that flow and channeling it toward safe ends—such as celebrating the Iron Range's mining history—that did not threaten the industrial communities of the Iron Range and their accumulated cultures. This double move, both encouraging tourism and also controlling it, lay behind most of the Iron Range's tourism efforts in the late twentieth century.[27]

Moving the interpretative center from an IRRRB plan to a completed project required significant funding. During the mid-1970s, the IRRRB slowly secured significant grants from federal, state, and local governments. Major funding came in 1974, when the federal Economic Development Administration pledged more than half a million dollars toward the project. The Upper Great Lakes Regional Commission paid $150,000 and the city of Chisholm, chosen as the site for the center because of its location near the center of the Mesabi Range, contributed $75,000. At the state level, the IRRRB put up the bulk of the funds for the project, contributing $780,000, while other state agencies, such as the Minnesota Department of Economic Development, the Minnesota Revolutionary Bicentennial Commission, and the Minnesota Resource Commission, contributed smaller amounts. As plans progressed in the early 1970s, IRRRB officials pushed for the center to open in time for the 1976 national bicentennial celebration. The bicentennial not only brought unprecedented attention to U.S. history in the mid-1970s, but it also made critical funding available for historical projects such as the Interpretative Center.[28]

With funding in place, construction of the Iron Range Interpretative Center in Chisholm began in August 1974. The 33,000-square-foot building, perched near the edge of an abandoned mine pit that was slowly filling with water to create an aquamarine lake ringed with red cliffs, was not completed until 1977. The building proved far more expensive than initially estimated, eventually costing $1.78 million to build. The center finally opened to the public on August 19, 1977, after almost ten years of planning and three years of construction.[29]

The historical narrative initially offered to visitors at the Iron Range Interpretative Center was crucial to the efforts to promote tourism and economic development on the Iron Range while simultaneously controlling outside interpretations of the region. Many Iron Range residents, including many of the planners at the IRRRB, were very conflicted about tourism. They acknowledged its importance in creating a mixed economy beyond mining, but they also feared its corrosive influence in a close-knit industrial region with an identity shaped by hostility to the outside influences that made life in a mining region so capricious. The compromise that all parties could agree on was that the center would focus on the region's mining history. By emphasizing the region's history of rich natural resources, the hard work required to transform those resources into valuable commodities, and the distinct local cultures created in and around the mines, the Iron Range Interpretative Center emerged as the focal point for a nostalgic history of the Iron Range that simultaneously celebrated an industrial past while moving forward into a postindustrial future.

The Iron Range Interpretative Center's historical narrative focused on five main categories: "geological and natural history," "historical and cultural interpretations," "the era of natural ores," "the taconite age," and "today-tomorrow." IRRRB promotional literature promised a historical narrative that "takes the visitor through a time capsule of history as the area is transformed from a region of dense wilderness to a highly industrialized society in the span of less than one hundred years." Implicit in this narrative was a story of human progress through technology and labor. Beginning with the natural resources of geology, the narrative demonstrated how industry transformed the region's rocks into iron ore through hard work and mechanical ingenuity. When depleted resources—the natural ore, in this case—threatened the progressive trajectory of the story, a new technological savior appeared in the form of taconite. Taconite, and promises of future ingenuity, launched the Iron Range into "today-tomorrow." This narrative was not just conveyed through exhibit scripts. The design and layout of the building also contributed to the narrative of progressive ascent from geology to a postindustrial future. Visitors entered on the ground floor—near the natural resources that spawned the Iron Range—and moved up and out as they progressed through the facility. The experience culminated with the view from a glass-enclosed walkway cantilevered out over the mine pit below. Standing in the cool, steel-and-glass walkway, looking out over the abandoned mine, visitors were subtly reminded about a modern future for the Iron Range that acknowledged and transcended the work of the past.[30]

Before it opened, the Iron Range Interpretative Center was embroiled in controversy. Built to span the divide between tourism promotion and the Iron Range's mining economy, it soon became a space onto which Iron Range residents could project their fears about an uncertain future. The first major controversy concerned the role that the mining companies would have in the new center, specifically whether they would donate money to the building and its exhibits. The catalyst for the controversy was a funding shortfall. The IRRRB simply did not have enough money to pay the large fee owed to consultant Joseph Wetzel for designing and fabricating the center's exhibits. Wetzel, a Boston-based exhibit designer, was hired by the IRRRB because of his reputation for high-quality design and planners believed he brought legitimacy to the upstart museum. During initial planning meetings Wetzel pushed the IRRRB to bring in additional consultants, whom he insisted would bring authority to the museum. He bragged that the "executive producer of Sesame Street" would be involved with the design and "these people bring a very fresh and particular perspective to the planning of a museum exhibit program." Although Wetzel and the consultants were hired to create the museum's exhibits, the IRRRB could not afford to pay the almost seven hundred thousand–dollar bill owed to Wetzel. The IRRRB hoped that the mining industry would cover the shortfall, but additional funds from the mining companies were not assured. Thus the center bogged down in a funding controversy before it opened to the public.[31]

The mining companies initially refused to contribute the requested four hundred thousand dollars toward completing the museum. Their refusal incensed several members of the IRRRB, who noted that the companies spent hundreds of thousands of dollars each year for advertising but would not pay for a museum about their industry. Trying to attract needed funding, politicians suggested that the center's portrayal of the Iron Range's history should be determined by who paid for the museum. Groups that contributed would be remembered favorably in the museum, while the stingy would be abused in the center's historical narrative. IRRRB member and Iron Range legislator George Perpich fumed over the mining companies' failure to invest in the Iron Range's history: "you know I've never heard a mining company donating much of anything . . . If a town is burning up, they'll send out a water truck, grudgingly." Perpich suggested that if the mining companies would not pay for the exhibit design, the museum's narrative should focus on mining accidents and generally "show [the mining companies] in an unfavorable light." He also proposed that the museum pass out flyers arguing that the steel industry did not contribute to the community in equal proportion to other Minnesota industries. Another IRRRB member suggested that the USWA be invited to contribute to the museum, in the hopes that adding a pointed labor perspective to the museum's script would persuade the mining companies to pay for a countering message. For their part, the mining companies were actively involved in the construction of a large statue of an iron miner immediately across from the interpretative center site in Chisholm. The statue was built from material donated by the mining companies and the firms argued that it was too much to support both the statue and the interpretative center. Although some envisioned the center as the stepping stone toward a bright postindustrial future, the museum quickly bogged down in the messy finances and labor–capital hostility of the industrial-era Iron Range.[32]

Deep divisions within the IRRRB exacerbated early funding problems, as the IRRRB squabbled over whether the center was a good use of the taconite tax money entrusted to the board. During a heated meeting on March 3, 1976, the ongoing problem of funding the exhibit design reached a boiling point. An additional five hundred thousand dollars was needed to pay for exhibit construction, but this amount was close to the sum total of the money available in the IRRRB's accounts. When one board member moved to grant the money to the center, another member threatened to leave the room, leaving the board without a quorum and thus unable to vote on the motion. The member explained his extreme opposition to the center, saying that it "hasn't been one of my all-time favorites from day one" and that spending all of the IRRRB's money on exhibit construction would be "an inconscienable [sic] act on the part of the IRRRB." The financial shortfall also caused rancor between the IRRRB and the center's staff. During the summer of 1976, the IRRRB suggested that center director Robert Scott deserved some of the blame. Board members felt that he was responsible for the failure to attract mining company financial commitments and during the June 23 meeting they recommended that he be removed as director. Eventually, the IRRRB resolved the funding problem by paying a greatly reduced amount—$150,000—for the exhibit design. The IRRRB stipulated that "none of this money would be used to construct exhibits which deal directly with the mining industry." This intense debate over funding, which erupted before the center opened, foreshadowed future debates over funding and finances once it was up and running.[33]

Indeed, after the center opened in 1977 it remained a focus for controversy, especially surrounding its tenuous finances and inability to generate enough revenue to meet expenses. At the core of this debate over finances were conflicting ideas about the center's mission. Was it supposed to operate as a profitable tourist attraction? Or should it fulfill an important public function, conveying the Iron Range's history to visitors, with the acknowledgment that it might never break even and would be a long-term expense for the IRRRB's finances? These questions remained unresolved throughout the late 1970s and early 1980s, leaving proponents on all sides disappointed. The IRRRB, for example, consistently pushed the center to break even financially and fretted over its drain on the agency's finances. Only a few years after the center opened, the IRRRB worried about its high costs and low attendance figures. During 1979 meetings, the board considered closing the facility during the winter to save money on salaries and heating costs. Conversely, staff insisted that the center's success could not be defined in purely financial terms. When IRRRB members urged the director to charge entry fees to offset expenses, the director vigorously protested: "it has always been the intention of the program that it would be free of charge . . . to allow as many people to come and learn the history of the Iron Range." Throughout its first years of operation, the center was the site of conflicting ideas about the nature and goals of historical tourism and historical memory on the Iron Range.[34]

Early attendance figures were disappointing and fueled the IRRRB's fears that the center would never operate without significant government funding. Built to accommodate 225,000 visitors a year, attendance during the first seasons was less than 100,000. By 1980, outside reviewers noted the harsh reality that the center was attracting only 70,000 to 80,000 visitors per year and was unlikely to dramatically increase its draw. The review urged the IRRRB to deal with the difficult decision of where the center should go in the future. It said that the center could reasonably expect to draw, at most, 150,000 visitors per year and concluded: "the IRRRB must face the fact that the Interpretative Center may never be financially self-supporting." To deal with this prospect, the center was urged to cut back on staff, perhaps even "contract out" all operations, and stop further development at the site. After setting unrealistically high attendance goals—necessary, perhaps, to justify the center as the base for a tourist economy—it was seen as a failure from its first years of operation.[35]

The IRRRB did not halt development at the center, however. In response to lackluster attendance in the late 1970s, the IRRRB instead expanded the center significantly, adding five additional components: a Hall of Geology, an ethnic crafts demonstration area, a conference or performance center, a historic cable car, and an Iron Range Research Library.[36] By the end of the decade, then, the center was moving in many different directions, trying to connect with any possible audience in the hope of drawing visitors and their money. Among the most significant expansions was the Iron Range Research Library—later renamed the Iron Range Research Center (IRRC)—a historical research library and archive that marked the IRRRB's most significant attempt to use history both as a tool for development and as a conservative force moderating the effect of change on the Iron Range.

The IRRC was first conceived in 1977 when the IRRRB decided to expand the newly constructed center. Construction of a research library began in 1978 and the facility was opened in 1980. As was the case with the interpretative center, proponents of the IRRC hoped that history and, in this case, scholarly historical study would contribute to economic development efforts. According to the IRRC's proponents, "not only will the Research Center work toward the goal of ensuring that the social, cultural, economic, political and religious history of the region is preserved, plans call for the facility to become the data base for research materials for developing programs of economic development, urban renewal and industrial diversification." Hoping that history would offer a "database" to guide future economic development efforts, the IRRRB and backers of the IRRC dreamed that the Iron Range could move boldly forward into a postindustrial future while retaining the lessons of the past.[37]

The IRRC was intended to lend scholarly credence to the newly created museum. Iron Range planners realized that creating historical tourism sites without a background of scholarly history ran the risk of building what were little more than history-themed tourist traps. But creating an archive to serve as the primary site for preserving the Iron Range's history created new problems as various local historical societies, the Minnesota Historical Society, and individual Iron Range residents battled over who had the right to preserve their memories and how their past should be interpreted. The conflict over control of the Iron Range's history played out—as conflicts over historical memory often do—as an institutional fight over boxes on shelves. Although the IRRC was built to house the Iron Range's historical archives, it needed material to fill the archives. By the early 1980s, local historical societies scattered across the Iron Range had been collecting small archives for years. These local historical societies held the collections that the IRRC would need in order to legitimately present itself as a viable historical research center. Local historical societies were hesitant to turn over to the IRRC their carefully collected and maintained archives, however, without assurance that they would retain some control over the use of those archives. Another complication was the initial plan to have the Minnesota Historical Society (MHS), the state's main historical archive and research organization based in St. Paul, control the IRRC. The combination of outside control and moving to a new facility created hard feelings. Many Iron Rangers who had a long involvement with local history through the historical societies felt that the IRRC was taking over their domain and ignoring their input. This conflict was phrased as a battle between the Iron Range and outsiders. As a representative from one Iron Range historical society put it, "This structure [the IRRC] belongs to the people of the Iron Range, not people from the big city. And the people from the Iron Range want to have some role in that research center." Local Iron Range historians were particularly upset that officials from the MHS—whom they associated with the Twin Cities—would be controlling the facility. During a 1981 IRRRB meeting, for example, representatives from the Iron Range historical societies criticized the MHS's handling of the archive and expressed fears that the Iron Range's history would no longer be controlled by Iron Range residents. As the meeting grew more heated, one IRRRB member insisted that the MHS did not respect the Iron Range Interpretative Center and the attached IRRC: "[MHS] didn't want to get involved when [the IRRC] was started . . . In fact, they said it was going to flop. That is no way to interpret history. It's going to be a carnival. It's going to be Disneyland." The board member then insisted that the Iron Range was "the most important area of our state" and even "the most important area of our country because without the Iron Range the rest of this country would not have been developed. The steel from our mines made this country grow—put us in an industrial revolution which made the U.S. a leader in the world." The conflict over who would control the IRRC and thus the Iron Range's historical memory revealed profound differences over the public presentation of history for tourist consumption on the Iron Range. IRRRB members and tourism planners recognized the importance of promoting historical tourism yet wanted to retain complete control over its history.[38]

While the Iron Range Interpretative Center was a focal point for historical tourism on the Iron Range during the 1970s and 1980s, several smaller heritage tourism projects were launched on the Iron Range during these years. These projects, ranging from statues to reclaimed mines to bold efforts at restoring an abandoned town, were driven both by economic necessity and the desire to use history to create a brighter future on the Iron Range.

Mining companies built a large statue of an iron-ore miner to commemorate the industry in the late 1970s. The statue, called *The Emergence of Man through Steel,* was built in 1976 as part of the national bicentennial celebrations. A local sculptor was hired after he promised to use rocks from Minnesota's three iron-ore ranges in the base of the statue and sculpt a large representational figure of an iron miner. The seventy-foot-tall statue featured a highly realistic miner, described by the sculptor as "projecting strength, humility and weariness." Local boosters believed the large statue pushed back against the tide of abstract modern art, with one local official describing it as "a victory of realism in art over the abstract."[39] These artistic critiques served as proxies for broader debates about corporate power and human labor's value within the increasingly

A statue celebrating the taconite industry, known as Rocky Taconite, was built in Silver Bay, Minnesota, in 1964. Minnesota Historical Society Collection, ML3.9 SB r4.

The Ironman Memorial outside Chisholm, Minnesota, was one of several efforts to commemorate the Iron Range's mining history during the 1970s and 1980s. Photograph by Peter J. Markham, Loretto, Minnesota.

abstract corporate imaginations of late-twentieth-century global conglomerates.

The Iron Range was not alone in funneling debates about deindustrialization through statues. As art historian Kirk Savage notes, multinational corporations were strong proponents of abstract sculpture throughout the 1960s and 1970s. U.S. Steel commissioned the critically regarded sculptor Richard Serra to sculpt a major work for its Pittsburgh corporate headquarters in the 1980s. As the steel industry declined dramatically during ensuing years, Serra's abstract sculpture came to symbolize the steel industry's turn away from its core business of making steel and foolish ventures into corporate finance and acquisitions. In Youngstown, Ohio, as well, public statues became a focal point for community debates about the steel industry in the era of deindustrialization. Youngstown hired sculptor George Segal to create a statue commemorating the steel industry in the late 1970s. Segal's sculpture, *The Steelmakers,* featured representations of two steelworkers and was placed prominently in front of a downtown department store. When Youngstown's steel industry collapsed suddenly in the late 1970s, the statue, which cost the city $150,000, became a symbol of loss for the devastated city. Vandals uprooted one bronze worker and placed him outside town as though the statue was hitchhiking out of Youngstown.[40]

Another effort to use mining history to promote tourism was the Soudan Underground Mine park in the Vermilion Range north of the Mesabi. When underground iron-ore mining on the Vermilion Range declined in the 1960s, local planners looked to historical tourism as an economic remedy for the abandoned underground mines. The Tower Soudan mine in Tower, Minnesota, became just such a tourism site during the 1970s. The Soudan mine was closed in 1962 as the taconite boom made natural ore obsolete. After the mine closed, a local development council urged the mine's owner, U.S. Steel, to turn the mine over to the state of Minnesota to create a park. Local boosters were hopeful that a mine park "will spark a new wave of tourism in their area in the coming years and many feel it will provide a lasting boost to their economy." U.S. Steel eventually turned the mine property over to the state and by the early 1980s the abandoned mine was folded into historical tourism efforts on the Iron Range. Specifically, the mine was opened for tours and the site became a state park. Tours of the abandoned underground mine were intended to both promote tourism in the Vermilion Range area and introduce visitors to the esoteric and quickly disappearing world of underground iron-ore mining. The mine's historical interpretation was guided by a 1982 consultants' report that encouraged the mine to use the deep shaft as a literal metaphor for "descending" into the region's mining history. Consultants recommended that the mine tour emphasize three distinct time periods from the mine's history: the late 1800s, the 1920s, and the 1960s, with an emphasis on authenticity in each section. Visitors would literally descend into the past as they traveled to different levels of the mine. They were to be "treated as if they are V.I.P.'s touring the mine, as foremen [tour guides] explain the processes of ore removal and problems being encountered." The consultants encouraged tour guides to let visitors "give advice, or even assist with a piece of work." Such interpretations likely fueled the worst fears of anti-tourism forces on the Iron Range, suggesting that historical interpretations in the mines would not be sympathetic to mine workers but instead would cast tourists as management officials.[41]

As the Soudan mine developed as a tourist site, the historical interpretation presented to visitors emphasized the technology of mining and the many innovative machines and techniques used in the mine to excavate iron ore from deep below the surface. A planning document described how this interpretation would be presented to visitors: "Locked in back caverns far away from the main shaft are the unchanged remnants of experiment after experiment in ways to remove the hard ore from the earth. To each remnant is tied a human story of astounding ingenuity, imagination and daring." The mine certainly contained fascinating machinery and innovative technologies, but exclusive emphasis on mining technology also may have been an attempt to cut off other, more critical interpretations of the mine. By focusing on mining technology, planners hoped that visitors would not leave with skeptical views of mine-working conditions, the long-term environmental impact of mining, or the processes of capitalism that had left the mine abandoned. The mine's historical planners explicitly stated that the focus on technology should be used to cut off critical interpretations of the abandoned mine: "it is imperative that this mine be seen in the context of advancing technology and developing society in general, and as an integral part of industrial change in America and the world." As the planners no doubt realized, tours of the abandoned mine were likely to leave visitors with more questions than answers. If not addressed immediately, many of those questions would turn to disturbing themes of industrial decline, community abandonment, and the processes of creative destruction at the heart of capitalism.[42]

Another Iron Range historical tourism project was the creation of the Hill Annex Mine State Park and historical site. The IRRRB took control of the closed mine, formerly an enormous open-pit natural ore mine outside Calumet, Minnesota, in 1979 from the Jones and Laughlin Steel Corporation. Although planners were eager to add a large open-pit mine to the IRRRB's tourism program, former miners on the board were skeptical of the mine's prospects as a tourist draw. Having worked in similar open-pit mines, these board members associated the pits with heavy, dangerous work and could not see how visitors could possibly interpret their former work sites as spaces of awe and wonder. Speaking critically about the Hill Annex Mine project, one board member exclaimed, "those of us who have been down in the mines, naturally, are not very excited about it." Former miners understood the mines in economic terms and many were suspicious of plans to spend large sums of money on abandoned mines just to promote tourism. A board member joked that if the IRRRB devoted any more money to the abandoned Hill Annex mine, "we will have a working mine." To their eyes, the mines were a productive resource and any money spent on them should be an investment in exchange for ore. Just as miners had clashed with tourism planners over recreational tourism, so too did they have very different visions of the meaning and significance of the abandoned mines.[43]

Calumet Historic Restoration Project

None of the proposed Iron Range historical tourism projects in the 1970s and 1980s were as audacious as the Calumet Historic Restoration Project. The plan, which aimed at turning a declining mining town into a perfect replica of a 1920s "frontier" settlement for historical tourism, illustrated the desire to use history to promote tourism and economic development. The plan, known as the Old Calumet Restoration Project (OCRP), proposed a wholesale reinvention of a declining mining town. In addition to the extensive physical work required to turn a 1970s town into a replica of an imagined 1920s community, the OCRP required substantial cognitive work, as it asked planners and potential funding agencies to see the small town not as a deindustrializing settlement, but as a historical resource ready to be tapped by a combination of planning and funding. Although the ambitious plan was never completed, a close look at it demonstrates how attempts to promote historical tourism on the Iron Range required difficult work to recast the modern mining region as a historic place.

The plan to restore "Old Calumet" was first noted in the early 1960s as part of a broader Iron Range Development Plan. The OCRP was fully described in the Iron Range Interpretative Program presented to the state legislature in April 1971. From its inception, the OCRP forged an alliance between Calumet's history—now described as a resource—and tourism-based economic development. This plan appeared straightforward at the time, but if viewed from a critical perspective it is clear that the project required a subtle recasting of the importance and meaning of history itself on the Iron Range. Through projects such as the OCRP and other efforts to promote historical tourism, past events and people were reconceived as a resource from which economic development could sprout.[44]

Inherent in the OCRP plan was the suggestion that Calumet's industrial decline was an inevitable—and perhaps even welcome—step toward the town's rebirth as a center for historical tourism. In an informal history of Calumet written as the restoration project was under consideration, a local booster portrayed Calumet's wholesale abandonment as an opportunity in disguise because the town was now ripe for reconstruction as a historical village: "Many of the buildings that are still standing are empty . . . Calumet is considered an original range mining town. An opportunity that . . . will never come again." Explaining the goals of the restoration project to the IRRRB in 1973, a planner described the project as an opportunity to "bring life . . . to a real dying, decaying community that is really just holding its own and breathing its last . . . this would be an opportunity to bring this community to life, rather than just an aesthetic museum, but really have the whole community a very definitely invested part of the past that can stand on its own two feet economically."[45]

The proposed project emphasized that historical tourism would preserve and honor the legacy of Iron Range residents: "The era of the early 1900s witnessed rapid development and extensive immigration. The story of how and why this development took place and who the people were, portrays a fascinating story . . . one that should never be lost." Although the description of Calumet as "an opportunity that . . . will never come again" is somewhat confusing—the town was, after all, slowly declining in population but clearly not going anywhere—efforts to justify the project hint at the tensions surrounding tourism based on historical reconstruction. By relegating the town's industrial past to the realm of history and nostalgia, supporters of the reconstruction plan were trying to simultaneously appease the living residents of Calumet, who had built their town and their lives from the mining industry and saw it as a vital matter of present concern rather than past nostalgia, while moving the town toward a viable economic future in a postmining world.[46]

In addition to questions about the motivation behind the Old Calumet Reconstruction Project there was another, seemingly more straightforward question: how exactly would the town of Calumet be "restored" to its 1920s condition? Planning reports describe just how much construction would be required to re-create the town as it existed in the 1920s. According to planners, "The first phase [of the project] should include absolute minimums or just enough to 'open the doors to the project.' Needed are enough buildings for people to visit, interesting items they may purchase, a convenient place to park and exhibits to observe and participate in to make the experience most enjoyable." The plan called for turning a garage into an "old fire hall" where "old fire equipment would be displayed with audio equipment relating stories of the serious fires that ravaged Main Street buildings in the early days." An operating pool hall presented a problem, however, as it was "not itself of historic importance" and its appearance was "not in keeping with adjoining historic buildings." The solution proposed was to use the pool hall building as "a center for renting bicycles" or "arranging for horse and buggy rides or other forms of transportation." The plan further called for the main street to be closed to vehicles at times with "piles of logs and beer barrels" that would "suggest the ruggedness of those early years on the Range." Important buildings that had been demolished could be cheaply framed "to suggest the appearance of important landmarks no longer standing," while keeping costs low by creating only "skeleton frames of their likeness . . . using old lumber." Although the scope of the initial plan, or phase one, was limited because of funding considerations, plans for an ambitious phase two included possibilities such as a printing shop that "could sell papers to tourists with their names in headlines, like 'John Doe Arrested for Bootlegging in Old Calumet,'" a sign maker who would sell signs and other crafts, an ice-cream parlor, a replica jail, a replica livery stable, and a replica blacksmith shop. The optimism of planners was reflected in their suggestions for long-range items such as an airstrip to bring in tourists, a "winter recreation area," and the subdivision of adjacent open land for a "tourist center" and for "motel development and other tourist services that would not be located within the historic district due to strict district guidelines." Re-creating Calumet as it existed in the 1920s involved far more than stripping away fifty years of progress. It would mean literally rebuilding the town.[47]

Thus, the plans for the OCRP reveal a fundamental contradiction in the use of history as an economic development strategy: Old Calumet, which supposedly existed before present-day Calumet, had to be constructed out of the 1970s town. As the plan moved forward, supporters performed complicated mental gymnastics to justify the significant physical construction work required to bring back the past. The key to resolving this contradiction was to emphasize historical authenticity, which, in theory, would allow even a newly constructed Old Calumet to be historically accurate. The attention to historical authenticity arose most palpably when defending the plan against the possibility that Old Calumet would be nothing more than a tourist trap. Considering the tremendous work required to restore Calumet to its 1920s state, even supporters of the project admitted that Old Calumet would not be a restoration so much as a new, historically themed entity. Yet a historical tourism site that relied on an "inauthentic" history was just a tourist trap. For Old Calumet to work, one newspaper editorial argued, "painstaking research and dedication will be needed. Authenticity is paramount in any restoration, for without authenticity, there is nothing. While Calumet, returned to its original state, will also draw tourists, it is much more important that it be a completely valid example of the early iron mining towns." Planners also argued that strict oversight of building design and "making certain that merchandise sold is in keeping with a goal of authenticity" would prevent the OCRP from becoming a historical tourist trap. At the earliest stages, the planners of the OCRP turned to authenticity as a means to justify their construction of a new town that re-created an older version of Calumet.[48]

Yet agreement that historical authenticity could prevent the OCRP from becoming a tourist trap left a key question unanswered: what did historical authenticity mean? Caught between a desire to maintain historical authenticity and the need to create a new Old Calumet out of 1970s Calumet, planners relied on odd tactics in their quest for authenticity. The most obvious step in this process was removing certain objects from Calumet buildings that were not deemed authentic by 1920s standards. Some buildings would require substantial work, such as relocating a telephone booth, adding new awnings and windows, and removing "antennas, and . . . modern day signs." A note on a restoration plan includes the following instructions for restoring the town's houses: "To restore the houses in Calumet they must: remove T.V. antenna; put wood trim over the windows; put wood siding and trim over the house; put wood boards over the concrete block on bottom." The status of the town's abandoned structures was more complicated, as they were perhaps the most authentic buildings, but their dilapidated state worked against the project's overall goal of promoting tourism. Thus, in one version of the restoration plan, the author argues that planners should work to turn the town's "dilapidated and abandoned" buildings into a re-creation with "the look of the 1920s." As these ambitious plans make clear, creating historical authenticity was an ongoing process that required constant and difficult work in the present to make an imagined past accessible.[49]

The realities of 1970s Calumet also forced planners to contemplate some re-creations that had little or no connection to the town's history. To reconstruct the town as it might have existed in the 1920s, planners suggested constructing entirely new buildings. Creating these "ghost buildings along Main Street" would mean constructing new buildings intended to be abandoned, because the originals had already been torn down. In a plan that called for new buildings intended to be abandoned and for refurbishing abandoned but still standing buildings to keep them from looking abandoned, one can sense the profoundly arbitrary nature of historical authenticity in Old Calumet. Although Calumet's historic district had only one church, planners argued that "if a church building considered . . . to be of historical significance becomes available, it could be moved to 'Old Calumet' from another range location." The tightrope of historical authenticity required planners to maintain an uneasy balance between authentic structures—such as churches that had really existed in the past—and the demands of a tourist center, which included bringing that church to a new location to make it more accessible to potential tourists. From an on-the-ground perspective, maintaining historical authenticity was an arbitrary venture that deemed some difficult ventures authentic, such as moving an old church to the town from another site, while determining that concrete block foundations were inauthentic. Historical authenticity, from this angle, did not so much resolve the contradictions inherent in constructing a new Old Calumet as it offered a unifying logic for the ultimately arbitrary decisions of planners.[50]

From a financial standpoint, the creation of Old Calumet was daunting. The 1974 plan estimated overall costs at $1.1 million, with even a limited first phase costing $568,400. The total valuation for the village of Calumet at the time was only $242,000. Planners were apparently unfazed that re-creating the town as it once was would cost more than four times the 1974 value of the town. As was demonstrated by several historical tourism projects on the Iron Range, returning to the past was often prohibitively expensive in the present.[51]

By the end of the 1970s, the IRRRB had committed to heritage tourism as a key component of its economic development plan, moving forward on numerous projects that sought to simultaneously celebrate the Iron Range's history of industrial iron-ore mining while drawing tourists and their money to the region. Although several of these projects had difficulty meeting their financial goals, planners envisioned their continued success in the years to come.

The 1980s Steel Crisis

The many different historical tourism projects under way on the Iron Range were upended in the early 1980s by the recession that decimated much of the U.S. steel industry. After decades at the center of the U.S. economy, the steel industry collapsed during the early 1980s as the bloated and outmoded steel companies succumbed to sharply increasing global competition, decreasing demand for steel in the developed world, and changing technologies.[52]

The iron-ore mines of northeastern Minnesota were doubly affected by the steel crisis. Many of the steel corporations that owned the taconite mines collapsed during the crisis and a growing stream of cheap foreign ore undercut the Mesabi Range's position as a key source of iron ore for the global steel industry. Although foreign iron ore began entering the American market in the postwar decades, by the last quarter of the twentieth century this stream had grown to a flood of iron ore that increasingly displaced the Mesabi Range as a primary producer of raw material for the world's steel industry. South American iron ore had challenged Lake Superior ores for many decades in the middle of the twentieth century, but by the 1980s advances in shipping technology and the full development of Latin American mines meant that South American ore could undersell Mesabi Range ore even at the Great Lakes steel mills. For example, Great Lakes steelmakers could purchase iron ore from Brazil at a 20 percent discount off Mesabi Range ore by the early 1980s. The rapid growth of foreign ore supplies and increasing global competition, a development that spanned the twentieth century, hit the Iron Range with full force by the early 1980s.[53]

The early 1980s steel crisis created the American Rust Belt. Cities such as Pittsburgh, Youngstown, Ohio, and Johnstown, Pennsylvania, drowned under waves of mill closures and wholesale community abandonment. The mining communities of the Iron Range, still predominantly one-industry towns despite several decades of economic development efforts, were similarly affected by the industry's collapse. Across the Iron Range in the 1980s, the steel crisis radically cut back work in the mines, drained cities of population, and even caused many residents to rethink their attitudes toward the economic importance of the region within the national and world economy. The most direct effect of the steel crisis was to dramatically cut back work opportunities in the taconite mines. Between 1980 and 1988, the number of workers employed in mining on the Mesabi Range dropped by almost 60 percent, from 14,000 to 5,500. By 1982, many Iron Range towns had staggeringly high unemployment levels. It was not uncommon to find 25 percent unemployment throughout the region and Hibbing's mayor guessed that unemployment "might be as high as 50 to 60 percent" in many Iron Range towns. A labor survey in the Babbitt and Embarrass area in April 1983 revealed the depth of Iron Range joblessness amid the steel crisis. Only 26 percent of respondents were employed, while 36 percent identified themselves as "unemployed and seeking employment" and another 36 percent claimed to be "unemployed and not seeking employment." While these grim statistics suggested widespread unemployment, out-of-work miners responded differently to the crisis depending on their age and work experience. One miner who lost his job in 1985 said that there was a distinct difference in response to the mine shutdown depending on the worker's age. He claimed that friends who began working in the mines right out of high school in the late 1950s and early 1960s refused to move on and hung around the area hoping to be called back. Those workers who were somewhat younger were more likely to move away, to places such as Utah, Colorado, and the Twin Cities.[54]

Two miners stand near a train car carrying a last load of iron ore from a mine in October 1983. Minnesota Discovery Center Collections, M. A. Hanna Company Records.

In the mining towns, the effects of the steel crisis soon rippled out of the mines to rock the social foundations of the communities. Overall, the recession of the early 1980s led to a 5.5 percent decline in the Iron Range's population between 1980 and 1988. It was hard to maintain the continuities of social life amid the deep economic disruption. One resident remembered the trauma of being in school during the early 1980s, where "during the first year after Butler [Taconite] closed, kids would just disappear from . . . class, each moving with their family as their parents tried to find work."[55] For those Iron Rangers who abandoned the region during the early 1980s, the nationwide recession meant that they had few opportunities in other parts of the country. For example, a 1982 *New York Times* report on increasing homelessness throughout the United States featured the story of the Thom family from Ely, Minnesota. James Thom owned his own home and a gas station in Ely but the poor regional economy led to the loss of his business and home. Thom, along with his wife and two children, were living in a school bus in Denver, hoping to make it to Tucson, Arizona, and the jobs they believed were available in the Sunbelt.[56]

Although not as dramatic as the Thom family's plight, the stories of other miners laid off during the 1980s steel crisis reveal how the economic downturn cascaded through the lives of Iron Range residents. After being laid off from Hanna Mining Company in the 1980s, Marsha Benolken did not know what other jobs she might find. Someone at the unemployment office told her there was public money available to return to school, so she used assistance to return to community college. Although she had previously attended the community college for three weeks before dropping out, she found that she now enjoyed school because it was "something different" from mining. After two quarters, though, she realized that she did not want to pursue secretarial work so she quit again and applied to work at a small factory in Hibbing, where she hoped to run a forklift. She described the toll that the layoffs took on her family's budget, which relied on two incomes:

> It was hard. When you have two incomes, you have bills or whatever for two incomes and when you go to one income people say, "oh luckily he has a job," but with the house and the car payment there's not much left over. Right away I thought of how are we going to make it? And you're used to living a certain way, so you really have to adjust. It's hard.

When asked specifically what she had to give up, Benolken said she began buying generic groceries and no longer went out. She and her family went camping instead. By the early 1980s, many families were relying on dual incomes to make ends meet. Benolken's story highlights the growing precariousness of many families' budgets by the late twentieth century. Ultimately, Benolken believed it was unlikely that she would ever again have a job that paid as well as the mining job she lost. Knowing that the unions kept pay high, she was resigned to a future of lower-paid employment and diminished expectations.[57]

The 1980s steel crisis led to a reevaluation of the Iron Range's economic role in the United States. Many residents interpreted the crisis as evidence that their work as industrial miners was no longer central to the nation's economy and culture. For a generation of residents who were raised on folk stories about the importance of the iron ore beneath the Iron Range—stories that inevitably focused on steel's role in winning both world wars—the shutdowns came as a psychic whiplash. Perhaps the Iron Range was not at the heart of the American, and thus the world's, economy. And, if this was the case, then what was the area but a provincial outpost in the north woods filled with abandoned mines? Pat McGauley, director of the IRRRB in 1982, expressed how the steel crisis changed many residents' attitudes toward the region's future:

> When things have been slow in the past, people have recog-
> nized that since the beginning of the ore mining, there have
> been good times and bad times, and that you just ride out
> the bad times. But people always thought, "We've got the ore.
> They've got to come here to get our ore." Now people have
> realized that there's iron ore all over the world—better ore,
> cheaper ore . . . They're aware that there are many things be-
> yond our control, and that they don't have to come for our ore.[58]

For laid-off miner Bill Cook, recognition of the Iron Range's precarious future was expressed as a desire to see his children avoid mine work. Asked what he would tell his children if they said they wanted to work in the iron-ore mines, he replied, "I don't think it's a good idea. Hopefully my kids will have enough intelligence to get a degree first."[59] The steel crisis devastated the Iron Range's economy and further contributed to the sense that the region was an anachronistic holdout from a quickly vanishing industrial past. Efforts to promote industrial heritage during this era were caught in a bind: by promoting heritage they often fueled the belief that the Iron Range belonged in the realm of history, but without heritage tourism, there were few routes for economic development during the steel crisis.

Ironworld USA and the Ethnic Revival

With the 1980s steel crisis focusing the attention of Iron Range politicians and policy officials on economic relief and development, efforts to promote heritage tourism, including the Iron Range Interpretative Center, were quickly folded into an urgent regional economic development plan that put increased emphasis and money toward any project that might conceivably bring jobs to the Iron Range. As a former IRRRB member put it, this "second phase of tourism development" "abandoned all pretext of a program based on history, culture and education. The main emphasis was on jobs." Given the urgent need for government stimulus in the area, tourism projects in the early 1980s were undertaken with haste and "precluded the time consuming process of applying for federal, state and foundation money." Additionally, the new push for rapid public spending was spurred by the priorities of Governor Rudy Perpich, a dentist from Hibbing who made Iron Range relief a centerpiece of his administration. Several policy makers felt that the push for rapid public spending on Iron Range tourism and public relief projects meant that the IRRRB largely ignored planning analysis in formulating regional development priorities in the early 1980s. Although heritage tourism projects had always been asked to serve two masters—history and economic development—the flood of public money in response to the steel crisis meant that heritage tourism was being asked to do more economic development work than ever on the Iron Range.[60]

One important development resulting from increased public spending on heritage tourism in the 1980s was a significant expansion of the Iron Range Interpretative Center and a new focus on white ethnicity in the region's heritage sites. Even though it was only seven years old at the time, the Iron Range Interpretative Center was remodeled and expanded in 1985 and 1986 as part of a $7 million project intended to make the center "one of the most modern, dynamic and popular tourist attractions in the United States." Along with this optimistic rhetoric, the expanded center was intended to bring many more tourists to the Iron Range. After the expansion, IRRRB officials optimistically hoped that the center's annual attendance would reach 350,000 visitors. The figure was optimistic given that only 750,000 visitors had come to the center over its previous seven years. Renamed Ironworld USA, the updated center reopened in 1986. Advertisements accompanying the reopening invited viewers to "Imagine Ironworld" and see where "a phenomenon occurs." In many ways, the expanded museum raised the stakes for using industrial heritage to promote economic development on the Iron Range. Officials dedicated even more money to the project in the hopes that an enlarged museum might succeed where it had struggled before.[61]

In addition to enlarging the museum, the expansion of Ironworld during the 1980s marked a new emphasis on ethnicity as the focal point of Iron Range history. By celebrating white ethnicity on the Iron Range, Ironworld was following a broad trend in American culture at the time: the ethnic revival. As historian Matthew Frye Jacobson argues, post-1960s American culture was awash in the rhetoric and symbolism of white ethnicity. The emphasis on ethnic "roots" offers another example of how the idea of a rooted heritage permeated various aspects of American life in the last decades of the twentieth century. Ironworld's focus on ethnicity and ethnic heritage expanded in the early 1980s when it opened an Ethnic Arts Center, a seven-thousand-square-foot space for demonstrating ethnic dancing, crafts, and artifacts.[62]

Among the most prominent celebrations of ethnicity on the Iron Range during this period were the ethnic festivals held at Ironworld. Like many small towns across the nation, the Iron Range had a long tradition of vibrant communal celebrations, including festive Fourth of July events in many towns.[63] By the 1970s and 1980s, however, a new kind of festival appeared on the Iron Range that sprang not from communal celebration but from the imperatives of planned heritage tourism. In 1978, the Iron Range Interpretative Center held the first Minnesota Ethnic Days celebration. By 1980, more than forty thousand visitors attended the twelve-day festival. The celebration was organized around various ethnicities, with each featured on a different day. Five elements of each ethnicity were featured on the day devoted to that ethnic group: "performance of music and dance," "demonstration of ethnic food preparation and ethnic crafts," "displays of the finery of each group," "films of the native homeland," and "opportunities to taste the various ethnic foods." The Minnesota Ethnic Days celebration highlights how the ethnic revival on the Iron Range during the 1970s and 1980s emerged from the needs of heritage tourism promoters eager to play up any distinguishing feature of the region in their bid for tourists. As performed during the celebration, ethnicity was largely an arbitrary construction, drawing on those parts of the immigrant experience that could be most easily marketed to tourists. Hence, by the early 1980s the Iron Range Interpretative Center began hosting the Iron Range Country Hoedown, an "afternoon of entertainment [that] presents the traditions, music and dance of the Ozarks." There were few connections between the Iron Range and the Ozarks, but a "hoedown" had the appropriate veneer of rural authenticity to attract tourists.[64]

Food and ethnic cuisine also fit the pattern of authentic ethnic traditions and soon became central to the ethnic revival on the Iron Range. In 1981, volunteers at the Iron Range Interpretative Center published *The Old Country Cookbook,* bringing together many ethnic recipes from the Iron Range and also featuring historic photos of the region. At Ironworld, however, there was little depth to this emphasis on ethnic food. Visitors and tourists gained little understanding of how cuisines fit into social and cultural networks or why many immigrants felt it was so important to retain food traditions. Instead, as Peter Goin and Elizabeth Raymond note, Ironworld "maps the Mesabi as a kind of historical food court, where immigrant cultures are manifested as distinctive cuisines."[65] The turn toward white ethnicity as a source of heritage during the 1980s was another illustration of the complexities involved in promoting heritage as economic development.

Despite several expansions and rebranding efforts, Ironworld's financial problems continued. Throughout the 1980s, the museum was criticized for its underlying approach to economic development through heritage. A 1980 consultant's report challenged the IRRRB's strategy of developing an Iron Range tourism industry based on pass-through visitors. The report argued that visitors rarely had extra time on their vacations to stop in the Iron Range and limited pass-through stops brought little economic benefit. The report went even further: "the mistakes made by the IRRRB—and by most others that have tried to tie history in with tourism—is in assuming that in promoting history, one promotes tourism." Critiques such as these challenged not just Ironworld, but the entire logic of heritage tourism on the Iron Range.[66]

These tensions were never adequately resolved at Ironworld. Was Ironworld a museum meant to teach visitors about the past or was it, as its director said in 1980, "not a museum" but "a communications tool."[67] The competing imperatives of emphasizing history and promoting economic development via tourism remained in tension even as it became increasingly clear that heritage tourism could never produce the broad economic gains its promoters initially touted. IRRRB officials nonetheless continued to promote Iron Range tourism throughout the 1980s. For example, they developed a sophisticated advertising campaign to bolster Iron Range cultural programs. Titled "Iron Range Country," the campaign featured billboards, radio advertisements, and print advertisements urging tourists from the Twin Cities, Iowa, and Chicago to "Come North, Visit Iron Range Country" and "Do the Range . . . We dare you to see it all!" Ironworld remained a focus for these tensions about tourism well into the 1990s and the first years of the twenty-first century. The museum turned toward corporate funding in the 1990s as public financing for the facility was cut back. In an ironic twist on the 1970s battles over corporate funding of the museum, Iron Range businesses were asked to contribute $2,500 to join Ironworld's "Heritage Corporate Club." After a comprehensive review of the facility in 1994, the IRRRB recommended that Ironworld be turned over to a private firm to manage for the 1994 season before announcing any long-term plans. A Twin Cities company was hired to manage the facility and charged with "increasing revenue and decreasing operational costs through cost-effective management."[68] In other words, the facility was privatized. Perhaps the saddest setback in Ironworld's saga came when the facility began collapsing into the very ground it was built to celebrate. In 2004, it had to be closed because of extensive foundation settling as a result of mine-induced subsidence. Built on top of abandoned underground mines, many unmarked, the entire facility risked falling into the ground. The museum was reopened in 2006 only after "structural reinforcements." Sinking into the geology it celebrated, Ironworld remains a cautionary tale about the use of history to promote economic development.[69]

Giants Ridge and the "Bavarianization" of Biwabik

Giants Ridge ski and golf resort and the nearby town of Biwabik illustrate another way in which the Iron Range turned to tourism in the wake of the 1980s steel crisis. Unlike Ironworld, which was mired in conflicts over industrial heritage, Giants Ridge epitomized a full commitment to recreational tourism. The skiing and golf resort was developed to be a lavish north-woods retreat for wealthy vacationers from outside the region. Giants Ridge was more financially successful than many other tourism projects on the Iron Range, but it also raised questions about whether the Iron Range could retain its industrial roots while catering to tourists.

The ski hill at Giants Ridge dated back to the 1950s when a small group of Iron Range residents carved a ski run out of a wooded hillside near the town of Biwabik. Local residents who enjoyed down-hill skiing were tired of driving hundreds of miles to the nearest ski hill, so they built a rough-and-ready facility on the Iron Range. During the 1950s and 1960s, Giants Ridge was a working-class operation, with truck engines powering a tow rope that pulled skiers up the short hill. A modified snowmobile groomed the snow, which was an improvement over the original method of hauling snow to bare spots in buckets and burlap sacks. A chalet with ninety beds was built in the 1960s. Rates were cheap at the new lodge. Seventeen dollars paid for two nights of lodging, three meals, and lift tickets. The ski hill was emblematic of many small, locally run recreational facilities in northern Minnesota prior to the 1980s. Giants Ridge ran into financial difficulties in the 1970s, however, after several warm winters left the facility with little snow. By the late 1970s, its directors were trying to turn the hill over to a municipal authority because they could not keep the facility financially viable. When the steel crisis of the 1980s hit, many of the hill's employees and volunteers were convinced that it would go bankrupt. "Everyone thought that was the end of the ski hill," one director said, noting that "it was a sad time for all of us who had put in more than twenty years building this dream."[70]

In 1984, the IRRRB stepped in to save the bankrupt ski hill and turn it into a centerpiece of the agency's tourism development efforts. The IRRRB was desperate to develop any business that promised jobs in the context of the debilitating steel crisis. Gary Lamppa, then director of the IRRRB, claimed that tourism projects such as Giants Ridge were "the best long-term development, with the shortest lead time" available for the agency to fund. After purchasing the hill in 1984, the IRRRB sunk several million dollars into Giants Ridge to create a larger and modernized resort for downhill and cross-country skiing. Both the IRRRB's purchase of the ski hill and investing in its modernization were controversial decisions. The U.S. Economic Development Administration refused to provide grant money to redevelop the site because it worried that the resort would produce few jobs and that Midwestern skiing was unlikely to be successful in the long run. Yet the IRRRB argued that a ski resort was "a result of the realization that a shrinking local industry demands alternative economic growth if the Iron Range is to survive. It is a tangible step toward increasing tourism in the region for the betterment of all segments of the economy."[71]

After several years of disappointing attendance in the 1980s,

Giants Ridge expanded to a year-round resort with the addition of golf in the 1990s. Because the resort sat empty for half the year, its management saw golf as a logical addition to attract tourists during the warm-weather months. In 1994, the IRRRB invested several million dollars to build a new golf course and supporting facilities. By the late 1990s, investments in skiing and especially golf appeared to pay off in the form of national attention and growing receipts.[72]

Development of Giants Ridge as a ski and golf resort also affected the nearby town of Biwabik. The small town was one of the earliest settlements on the Mesabi Range and, like the other small towns of the Iron Range, it grew alongside the open-pit mines in the early twentieth century. Biwabik was similarly hard hit by the steel crisis of the 1980s. Yet, with the nearby development of Giants Ridge as a ski resort in the mid-1980s, it began an ambitious plan to "Bavarianize" itself to appeal to tourists visiting Giants Ridge. After deciding that the mining economy would never return to its former glory, town leaders hired a consultant to suggest a redesign for the town that would appeal to tourists. The consultant suggested that Biwabik take on a Bavarian Alps theme that would visually connect it to Giants Ridge. Storefronts in Biwabik were retrofitted with faux alpine faces and a variety of alpine-themed bars and restaurants opened. Older residents and historic preservationists questioned the town's Bavarianization, pointing out that neither the town nor its residents had significant roots in Germany or the Alps. Yet, like various historically themed development projects mentioned, the goal of Biwabik's Bavarianization was not to preserve an authentic history. Instead, it was meant to make the town attractive and immediately legible to tourists seeking a fun and entertaining skiing vacation. Biwabik has been successful in this regard. After the town's facelift, tourist literature described it as "a northwoods town with a Bavarian personality" and noted that "Biwabik . . . has rejected its grimy mining past to cultivate a sleeker image as an après-ski playground."[73]

Although historical preservationists questioned the authenticity of a Bavarian-themed town on the Iron Range, Biwabik's redesign as an alpine ski town was motivated by an attempt by worried local residents to adjust to the new realities of a service- and tourism-based economy while holding on to some remnants of the town's traditions. For instance, in 2005 Biwabik's mayor argued that the goal of redevelopment was to give young people—residents' children and grandchildren—a chance to stay in their hometown instead of feeling as though they had to leave to find other opportunities: "I want it known that twenty years from now my daughter wouldn't have to leave, she could stay." Biwabik's city planner also justified the Bavarian makeover: "you can't just focus on going out and recruiting new investors to your area. First you need to make your area as attractive as possible, with facilities that appeal to the people you want to locate here."[74]

Biwabik and Giants Ridge illustrate an extreme example of how Iron Range communities remade themselves at the end of the twentieth century to appeal to tourists from outside the region. In many ways, Giants Ridge has been among the most successful economic ventures on the modern Iron Range. By the early twenty-first century, Giants Ridge was pumping money into the region's economy. In 2013, IRRRB officials estimated that Giants Ridge welcomed more than a hundred thousand visitors to the region and brought $8 million of economic impact. Yet these economic changes also brought cultural transformations that threatened to relegate the region's industrial history to the distant past. By the first decade of the twenty-first century, Biwabik was "becoming a hot spot for condominiums and other real estate development." The lodge at Giants Ridge was also redeveloped to add "specialty retail and a gourmet coffee shop for condo owners and tourists." If Giants Ridge represented a postmining future for the Iron Range, many local residents were left to wonder who benefits and who loses on an Iron Range devoted to providing memorable tourist experiences to visitors from outside the region.[75]

Keweenaw National Historical Park

Although there were many efforts to promote industrial heritage and tourism on the postwar Iron Range, most were local, regional, or state-led affairs. There was little direct federal involvement. The Keweenaw Peninsula, located on Michigan's Upper Peninsula across Lake Superior from the North Shore and the Iron Range, offers a useful contrast from heritage planning on the Iron Range. Like the Iron Range, the Keweenaw Peninsula was a mining region—copper mining, in this case—that had declined greatly in the postwar decades. Unlike the Iron Range, the federal government stepped in on the Keweenaw to preserve the region's mining heritage, ultimately creating the Keweenaw National Historical Park, a national park dedicated to mining heritage. It is important to note that there were several important differences between the Iron Range and the Keweenaw Peninsula that make direct comparisons impossible. Open-pit iron-ore mining was fundamentally different from underground copper mining on the Keweenaw Peninsula. Also, copper mining had collapsed totally on the Keweenaw Peninsula by the late twentieth century, unlike the continuing presence of taconite mining on the Iron Range. Nonetheless, the experience of heritage preservation on the Keweenaw offers a valuable nearby comparison to the Iron Range, notably illustrating how federal involvement failed to resolve thorny problems of balancing heritage and economic development in a postmining region.

Located on a rocky thumb of land jutting north into Lake Superior from Michigan's Upper Peninsula, the Keweenaw Peninsula launched the first mineral boom in the Lake Superior region. American prospectors rushed to the Keweenaw in the 1840s following reports of rich veins of native copper. Copper mining soon expanded to an industrial scale and the Keweenaw produced huge volumes of copper for American industry until early in the twentieth century, when the region was surpassed by more profitable western copper mines. Copper mines on the Keweenaw held on until final closure in the 1960s. Efforts to preserve and interpret the Keweenaw's copper-mining industry began in the 1920s and 1930s, almost immediately after the industry passed its peak. Preservation efforts were limited in the middle of the twentieth century but picked up again in the 1970s. At this time, a local business began planning a large copper-mining heritage theme park named Coppertown USA—note the similarity to Ironworld USA—that would preserve historically significant buildings from the region's copper boom and attract tourists to the remote peninsula. Coppertown USA proposed an ambitious open-air park, but it never was able to expand beyond a small museum in Calumet, Michigan. Coppertown's supporters urged the National Park Service to create a national park on the Keweenaw, but this idea had little political support in Congress and did not move forward at the time. The concept of a national park devoted to copper mining on the Keweenaw Peninsula was revived in the 1980s. Local residents and industrial archaeologists worried that the region's historic structures were quickly decaying and urged the federal government to step in and preserve them. In 1988, the National Park Service published a report indicating that a national park was feasible on the Keweenaw Peninsula and outlining how such a park could function.[76]

Supporters of a national park on the Keweenaw focused mainly on the economic benefits that it would bring to the struggling region. For instance, the region's congressman said that a national park on the Keweenaw Peninsula would "create a new era of economic opportunity for that area by capitalizing on that historic past—by sharing a very important chapter in American history with hundreds of thousands of visitors each year." As on the Iron Range, many politicians and local businesses paid lip service to the intangible value of historic preservation but they focused primarily on the bottom-line benefits of tourism. Indeed, the professionals hired to manage heritage sites often worried that too much focus on economic development would inevitably disappoint local residents and detract from important but unprofitable preservation efforts. These warnings were often ignored, however, in the rush to bring federal park-service funding to a declining mining region.[77]

As on the Iron Range, supporters of industrial heritage on the Keweenaw Peninsula tried, often with great difficulty, to balance competing demands for historic preservation and economic development. The National Park Service (NPS) was drawn into these debates when it began planning a national park in the region. Trying to balance the need for preservation with demands for economic stimulus and the realities of limited federal budgets, the NPS proposed a novel organizational structure for the Keweenaw National Historical Park. In its initial report explaining how a park on the Keweenaw could be feasible, the NPS proposed a small-scale park that operated as a hybrid public–private enterprise. The NPS would own a small number of buildings deemed central to the region's copper-mining heritage. It would also partner with affiliated heritage sites in the region that were privately owned. NPS officials were conflicted over the new park's viability and its unusual organizational structure. Among the park's supporters was its first director, who was excited over the park's new organizational structure: "The new park is an excellent prototype for new parks in the next century. We hope to have this park develop as a true partnership park—one where a relatively limited infusion of federal money acts as a catalyst to encourage the communities and their governments to take actions which preserve their heritage." Yet higher-level NPS appointees worried that a park on the Keweenaw Peninsula was a prime illustration of "park-barrel politics," in which national parks were used as indirect vehicles for funneling public funds to pet projects in legislators' home districts. The director of the NPS at the time was opposed to creating a national park on the Keweenaw. "[Keweenaw's copper industry] was certainly regionally significant, probably statewide significant," the director later said, "but we didn't feel that it held the true national significance that should make it a national park site."[78]

Pushed by powerful supporters in Congress, Keweenaw National Historical Park became a federal park on October 27, 1992, when President George H. W. Bush signed a bill authorizing it. Many Keweenaw residents were excited about federal recognition and cultural respect for the peninsula's industrial history. Calumet's village comptroller said, "I'm just kind of numb. It really happened." A state politician who supported the park described it as the "major, major achievement" of his political career. A local pastor wrote that the national park meant that "we are recognized for the outstanding contribution the Keweenaw has made in the past . . . It means we are superlative! It means we are historic! It means we are distinct! It means we are a resource to the nation!" There were, however, rumblings of discontent from Keweenaw residents who worried that the new park would bring an influx of "outsiders" to the region. An Oklahoma woman who moved to the Keweenaw to open a bed-and-breakfast heard many local residents complaining about outsiders arriving in response to the new national park. On the Keweenaw, as on the Iron Range, heritage tourism was a double-edged sword. It promised to revive declining industrial regions while honoring the industrial labor that made the regions distinct. Yet the reality of catering to tourists meant subtle yet significant transformations in the economic base and cultural significance of the regions.[79]

Keweenaw National Historical Park was embroiled in political controversy almost immediately after its authorization. Although the president's signature created a national park, its annual funding was dependent on Congress and the park had numerous congressional opponents. A representative from Michigan's Lower Peninsula joked, "this thing is so far north you can only reach it by dog sled, even on the Fourth of July." More important, a newly powerful caucus of conservative Republicans in Congress began combing through the federal budget in the early 1990s seeking to cut federal spending. The new park was one of their first targets and Congress sharply cut the park's budget. Keweenaw National Historical Park requested $1 million of federal funding in its first year, but Congress cut that amount to $150,000. The only consolation for park staff and local supporters was that the park was not being targeted for its historical value, but rather was being used as a scapegoat for critiques of federal park spending. "Our park has the distinct pleasure of being born at a time of great turmoil in our country over such minor issues as taxes, spending and deficits," the park's director wrote. Criticism of a park on the Keweenaw reached a national audience several years later, when the *Wall Street Journal* published a scorching critique of the park as a wasteful boondoggle of federal spending. "The true purpose of the park," it said, "is to prop up the economy of a remote area on Michigan's Upper Peninsula." The article went on to describe the park as the "bleakest" park in the National Park Service, sarcastically describing its offerings to visitors as "a close-up look at crumbling commercial buildings, slag piles and a Superfund hazardous-waste site."[80]

The federal government's role on the Keweenaw, via the National Park Service, was much greater than it was on the Iron Range. Federal assistance brought much-needed funds to the region, but it also turned the Keweenaw's economic development projects into a national debate over "park-barrel politics." In contrast, conflicts over heritage tourism on the Iron Range were limited to local and state politics. Nonetheless, themes of continuity run between both cases. On the Keweenaw as on the Iron Range, local residents and heritage professionals struggled to balance contradictory missions: how to develop a postmining economy based on tourism while retaining the region's industrial character.

Conclusion

When asked in 2000 why he would not leave the Iron Range despite a lifetime of hard times, USWA Local 4108 president Jerry Fallos replied, "This is our life. This is our heritage."[81] Fallos's comment echoed the sentiments of many workers across the Iron Range and throughout postindustrial America at the dawn of the twenty-first century. Industry and industrial labor were important in these stories not because they contributed to a vital future, but because their long history had culminated in a rich patina of heritage—the sum total of lifetimes of hard work and tough towns built around the mines and mills. While the sentiment behind these stories is certainly real, this invocation of heritage should give us pause. As this chapter has shown, heritage on the Iron Range is both omnipresent and constructed; that is, constant talk about heritage should not obscure the term's creation in the imperatives—financial and emotional—of postwar planners struggling to balance postmining development with the fears and dreams of industrial miners anxious about losing their position at the center of American life in the twentieth century. This story is relevant for a broad swath of postindustrial America. Heritage, especially the broad-shouldered, social-realist industrial heritage common throughout the Rust Belt, was not a natural or inevitable result of history, but rather was a contingent and specific response to the political, cultural, and economic conditions of deindustrializing regions in the post–World War II decades.

It would be easy to dismiss the rhetoric of heritage as a harmless way of remembering American industry in its twilight, but the focus on heritage has also shaped the political response to deindustrialization in problematic ways. Specifically, industrial heritage as practiced in places such as Ironworld emphasized the local industrial community as the logical and natural site of identification and political activity. By foregrounding the local community, however, this version of heritage cut off possibilities for alternative languages and alternative collectivities that might have proven far more effective at countering, or at least ameliorating, the worst effects of deindustrialization. The emphasis on heritage as a tool for negotiating tensions in postindustrial America problematically foregrounds the local community at the expense of broader translocal and transnational affiliations. This was especially true for industrial workers whipsawed by changes in the global economy in the late twentieth century. The problem for many industrial workers was not that they had insufficient communities—indeed, the rich communities that supported industrial work have often lived on well past the industries they once served—but rather that many workers did not have a language appropriate for the scale of the problems confronting them. Historian Jefferson Cowie describes this dilemma: "the pull of place and community has been a powerful force in labor relations, but the limitations of local identity also create constraints on a more expansive notion of working-class politics in an era in which capital transcends boundaries with complete ease." Similarly, historian Steven High argues that labor historians are complicit in these narratives of working-class localism: "capital versus community" stories of deindustrialization "have left little room for larger trans-local modes of identification." In other words, the rich social worlds of working-class communities—an emphasis of historians and activists for many decades—may have foreclosed alternative lexicons for understanding the nature of global capital at the beginning of the twenty-first century.[82]

Another consequence of the emphasis on community as the site of authentic working-class experience was the absence of the nation in debates about deindustrialization. With the political rhetoric of community so prevalent in industrial towns, many observers interpreted American mine and mill shutdown stories through a purely local lens. Even when the federal government did step in, as happened on the Keweenaw Peninsula, its efforts were seen as local pork-barrel spending rather than a response to an urgent national problem. The overall narrative framing of deindustrialization, as Steven High has noted, became a "capital versus community" story that pitted ruthless international corporations against rooted working-class communities. Ironically, this community-based framework, favored by sympathetic journalists and academics, often ignored the language of nationalism used by laid-off workers to describe the causes of their plight. Reflecting on the 1980s steel crisis, for instance, one Iron Range miner blamed the shutdowns on the federal government's failure to help industry: "From here to Chicago to Gary to Detroit to Cleveland to Pittsburgh to Buffalo, the major industrial power base in this country is in deep trouble. You can go through small recessions and the normal highs and lows, but you don't have this kind of problem unless the government is being unresponsive." The contrast between the national rhetoric of many laid-off workers and the local, community-focused emphasis of many sympathetic observers perhaps goes some ways toward making sense of the simultaneous embrace of patriotic politics by many former industrial workers and acknowledgment of the harshly negative economic reality they faced in postindustrial America.[83]

Finally, heritage on the postindustrial Iron Range raises difficult questions for historians and heritage professionals. The turn toward heritage as an economic development strategy on the Iron Range in the 1970s and 1980s sprang from the best intentions of many different planners and heritage professionals. Heritage seemed to offer the best possible solution to the Iron Range's thorny problems at the time: a declining mining industry, isolation from population centers, and residents who were ambivalent about turning their home into a tourist destination. Given these constraints, perhaps heritage was the least bad option available.

But, no matter the alternatives, heritage was not without consequences. By promoting the Iron Range's industrial history, heritage also made the move away from industry seem natural and inevitable. The concerns of miners, both active and former, were moved to the realm of nostalgia, where they could be safely handled in a way that had little effect on future plans for the area. In short, heritage depoliticized industrial decline on the Iron Range. Thus, heritage on the Iron Range and elsewhere offers a cautionary tale about the unintended but nonetheless very real consequences that follow from certain attempts to promote history.

CONCLUSION

Lessons for a Future Iron Range

Throughout the twentieth century, Minnesota's Iron Range struggled with decline and deindustrialization. The Iron Range was created in the late nineteenth century to exploit minerals needed for
the iron and steel industry. It rose and fell alongside that industry
over the decades, and it struggled with the inherently unsustainable nature of a resource-extraction industry. Yet the people of
the Iron Range, joined by allies throughout Minnesota and across
the nation, fought against decline using every tool and technique
available. The Iron Range did not accept the fate of dissolving into
a string of mining ghost towns. In this struggle against decline, the
Iron Range's history contains valuable insights into the processes
of industrial decline in the modern United States.

First, the Iron Range's struggle against decline offers a useful
vantage point for understanding the interconnected fates of industry and liberalism in the United States in the late twentieth century.

On the Iron Range, politicians committed themselves to active governmental involvement in the region's economic health, especially the iron-ore mining industry. Deindustrialization undercut the alliance between labor and liberalism in ways that neither DFL liberals nor Iron Range residents could anticipate. As globalization and automation reshaped the iron-ore mining industry, liberal politicians found that their policy tool kits contained few remedies for long-term industrial decline. They had surprisingly few answers for residents of the Iron Range facing shutdowns and layoffs. Conversely, deindustrialization revealed the pragmatic core of many working-class Iron Range residents' commitment to liberalism. Miners supported Minnesota's DFL liberals mainly because of the economic benefits liberalism offered to them. This alliance, however, was built on a foundation of industrial growth. When that growth ended, politicians and residents worked diligently to maintain their previous political relationships, but the alliance of industrial liberalism gradually eroded as the twentieth century wore on and industrial decline continued. The Iron Range has not followed the familiar pattern of industrial workers moving from left to right in the late twentieth century. Instead, the politics of the Iron Range suggests a more complicated transition away from midcentury liberalism based on industrial growth, even in those regions most deeply committed to it.

The Iron Range's response to industrial decline also reveals how the predominant policy response to deindustrialization in the United States, economic development policy, was enacted as a local response to national and global problems. Through the IRRRB, the Iron Range and the state of Minnesota made a concerted effort to ensure that the Iron Range would avoid the fate of many other industrial regions suffering from decline in the postwar era. In many respects, the Iron Range was unusually successful in fending off the blight of deindustrialization. It retained jobs—often at great public cost—and avoided the outcomes that befell smaller cities dependent on the steel industry such as Youngstown, Ohio, and Johnstown, Pennsylvania. The Iron Range thus offers an example of the possibilities inherent in vigorous economic development policy during the postwar era. If local and state governments carefully managed their resources and spent enough public money, it was indeed possible to keep industrial jobs alive and fend off the worst effects of deindustrialization. The Iron Range's success with economic development policy, however, also illustrates the limits of local economic development efforts in responding to what were ultimately global phenomena. Despite the earnest efforts of economic development professionals, it was ultimately impossible to reverse the iron-ore mining industry's increasing automation and globalization during the second half of the twentieth century.

It is impossible to separate deindustrialization and technological change in the postwar history of the Iron Range. More than any other factor, technical innovation in the mining industry displaced jobs throughout the twentieth century. Automation increasingly shifted work from human laborers to machines of increasing size and sophistication. The taconite industry offers the starkest example of how automation, and technological change in general, was a double-edged sword for the Iron Range. On the one hand, the taconite industry opened up vast new deposits for use as iron ore, likely prolonging the life of the Iron Range's mineral deposits by many years. On the other hand, the taconite industry's success meant that the existing natural ore mining industry became obsolete. Through technical innovation and the rhetoric of depletion, Edward W. Davis and other promoters of the taconite mining industry simultaneously developed a new industry and dismantled the older natural iron-ore mining business. While they touted taconite as a technological miracle to save a depressed Iron Range, scientists and engineers often ignored the destructive work that accompanied their creation. In a larger sense, both industrial regions and labor historians have yet to grapple fully with the complicated implications of technological change. In much economic development literature, for instance, there is still hope that "high-tech" miracles will revive sagging rural economies or that new communication technologies will erase inequalities of distance and capital. Additionally, historians of labor and industry often downplayed the central role of technological change and automation in reorganizing patterns of work throughout the twentieth century.

Finally, the Iron Range's attempt to promote cultural attractions and especially heritage tourism based on mining's history raises thorny questions about the role of history and heritage in depoliticizing industrial change. On the Iron Range, history became a vehicle for moving deindustrialization out of the realm of politics and into an apolitical realm of nostalgia. In museums such as Ironworld, mining's history was simultaneously celebrated and foreclosed as a possible future for the Iron Range. The heritage professionals and historians who created this romanticized story of the Iron Range were not malicious. They were usually driven by a desire to honor and celebrate the lives of hardworking immigrants and the rich communities they created in a harsh landscape. But historical interpretations have consequences, and one consequence of their interpretation was to depoliticize industrial change in northeastern Minnesota.

The Iron Range in the Twenty-first Century

Economic decline on the Iron Range and the fight against it did not stop in the early twenty-first century. Throughout the first decade, the Iron Range swung wildly between despair and optimistic predictions of growth. Although these experiences were well known to industrial regions by the early 2000s, the quickening pace of economic change on the Iron Range suggests that globalization has increased economic volatility in rural America and illustrates how local economies now respond almost instantly to global developments.

The first years of the new century began with yet another economic crisis. In May 2000, a major taconite producer, LTV Steel, announced that it would be closing its Minnesota mining operations because of the plant's obsolete technology and the declining quality of the ore body. LTV estimated that it would cost $500 million to modernize the plant to make it competitive in the iron-ore market. It chose instead to shutter the plant. More than one hundred LTV employees were immediately laid off and the remaining 1,400 workers were told they would lose their jobs gradually as operations wound down. LTV's president insisted that the closing resulted from impersonal market forces, not the work of the miners. The closing devastated the nearby town of Hoyt Lakes, which depended on the mine for jobs. As happened during downturns in the 1950s and 1980s, residents suggested that there was no future for the region and encouraged younger people to leave. One LTV employee described the advice he gave his two children in the wake of the closure: "I told them, get a good trade, but don't necessarily go to the mine . . . They all listened."[1]

State officials were sympathetic to the plight of the laid-off miners, but by the year 2000 many politicians seemed to acknowledge that mine closings were simply the unfortunate cost of doing business in the global steel industry. There was seemingly little that state government could do beyond offering temporary assistance to laid-off miners. Minnesota governor Jesse Ventura was sympathetic to the miners' plight, but ultimately believed such closings were just "the negative part of doing business." The LTV closing, which closely followed the growing economy of the 1990s, was a reminder for many on the Iron Range that even prosperous times could be quickly undercut by shutdowns and layoffs.[2]

Once the LTV crisis passed, however, the rest of the decade was a good one for the U.S. steel industry and for the Iron Range. With natural resource prices booming and newly emergent industrial powers such as China and India hungry for raw materials, the Iron Range rode a new wave of globalization in the steel industry. Business analysts noted that the early 2000s were a period of growth for the American steel industry, largely caused by consolidation in the industry, first among the integrated producers, then among raw material suppliers and minimills. The Iron Range profited handsomely from this boom, with several older mines reopening or expanding. Eveleth's EVTAC taconite mine reopened in 2003 when Cleveland-based mining conglomerate Cleveland-Cliffs and Laiwu Steel of China formed a new joint company, United Taconite LLC, to manage the mine. Laiwu Steel took 30 percent ownership over the Eveleth mine as part of a broad attempt by Chinese steel firms to meet quickly rising demand for iron ore in China. The reopening also required $2 million of public money from the state of Minnesota and the IRRRB. Iron Range miners often had mixed feelings about the expansion of mining during the 2000s. Miner Bill Matos, for example, described the EVTAC reopening as "bittersweet." When United Taconite had previously closed the mine, Matos was only two months away from retirement with a full pension. He now needed to work another five years at the reopened plant to reach retirement. Reopening formerly shuttered mines offered new jobs, but these jobs could not overcome several decades of decline and the lost wages and time of deindustrialization.[3]

The boom of the early 2000s made clear that the Iron Range's economy was deeply enmeshed in the global economy. As described in this book, Iron Range residents had always been aware of their region's connection to other mining regions around the world. In past decades, however, many residents and miners felt that the United States could protect itself from foreign competition if there was political will to enact trade barriers. This belief seemed to evaporate in the twenty-first century as Iron Range mining companies were increasingly owned by Asian steel producers. Many residents now saw that their local economy benefited from foreign demand. It was clear that the Iron Range's significance within the global steel industry had declined precipitously by 2006. In that year, the United States produced only 3 percent of the world's iron ore. Not only was the Iron Range supplying a smaller and smaller percentage of the world's iron-ore supply, but China supplanted the United States as the world's predominant steelmaking nation. Increasing Chinese demand for iron ore fueled the Iron Range's boom during these years. Little ore from the Iron Range was shipped directly to Chinese steel mills. Instead, ore from other nations, such as Canada, Australia, and Brazil, that typically competed with Iron Range ore in the U.S. market was now diverted to China. Iron Range ore filled in on the domestic market. In some cases, complicated business contracts sent Iron Range ore to Canada to replace Canadian iron ore that was sent to China. Minnesota politicians recognized China's new significance in the steel industry, and in 2005 the IRRRB's commissioner joined Minnesota's governor and several business leaders on a trade mission to China. The group connected Minnesota business interests with Chinese corporations and toured several Chinese steel mills. Iron Range residents were often conflicted about their economic revival thanks to Chinese demand. Miner Joe Strlekar, for instance, saw the situation as a reversal of the typical globalization story. "Instead of everything coming back into this country from China, it's good to have something going the other way," he told a reporter.[4]

The boom reached a peak in 2008 amid rapidly escalating mineral prices and an overheated global steel industry. In the spring of 2008, local media began reporting a broader "renaissance" on the Iron Range. Business analysts pointed to a dozen industrial projects under way in the area and more than $6 billion of investment money pouring into various projects. Once complete, these investments were predicted to create 1,400 full-time jobs. The high point of the 2008 expansion was the February announcement by U.S. Steel that it was expanding its Keetac taconite plant in Keewatin. This would be the first major expansion of taconite production on the Iron Range in two decades. The expansion, which was estimated to cost $300 million, would bring seventy-five new full-time jobs to the town and several hundred temporary construction jobs.[5]

As happened so many times during the postwar era, this boom did not last. The global financial crisis that began in 2007 quickly ended hopes of a revival for the Iron Range. The Keetac plant that was scheduled for expansion was idled completely in December

2008. Layoffs and shutdowns at all the remaining taconite plants quickly followed in late 2008 and early 2009. What some Iron Range residents found especially troubling about the 2009 slump was the speed with which the region's economy soured. The Iron Range went from expanding production to a deep slump in the course of several months in early 2009. Union official Mike Woods described the shutdowns as coming "almost overnight" and Sandy Layman, commissioner of the IRRRB, noted the "breathtaking speed" of the layoffs. The only silver lining in the downturn was that the economic pain was not limited to the Iron Range. Mountain Iron mayor Gary Skalko found solace in the fact that the Iron Range was "not losing people because there aren't that many places to move to—things are hard all over."[6] It is impossible to predict the future on the Iron Range, but it is clear that the swings between boom and bust are happening more quickly and with more force in the global economy of the twenty-first century.

Controversy over New Copper-Nickel Mining Ventures

Beyond the taconite industry, the early twenty-first century on Minnesota's Iron Range and throughout the entire Lake Superior region saw contentious debates over new mining ventures, including new copper-nickel mines, that raised old questions about the benefits and costs of sustaining the region's mining economy. As of this writing, many of these proposed mines are waiting for environmental permits and financing packages that may or may not materialize. And the ever-volatile global commodities markets could certainly upend plans as they did several times in the Iron Range's history. Yet what is striking about the debate over proposed new mines in the Lake Superior region is how they echo old arguments, such as Edward W. Davis's promotion of taconite as a technological savior for the Iron Range's beleaguered economy in the 1950s or worries that Reserve Mining Company's pollution of Lake Superior would have catastrophic consequences on human life.

Geologists and mining engineers were aware of copper and nickel deposits in the Lake Superior region as early as the 1950s. During extensive mineral exploration in the middle of the twentieth century, mining firms stumbled upon the Duluth Complex, a large deposit of copper, nickel, and platinum group minerals lying near the Mesabi and Vermilion ranges. Looking at the large, untapped deposits of copper, nickel, and other precious minerals, observers

223

CONCLUSION

claimed it was a matter of when, not if, new mining would start in northeast Minnesota. Yet the low grade of Duluth Complex ores, low commodity prices, and environmental regulations limited development of copper-nickel mining in northeast Minnesota for several decades. International mining firms returned to the region at the turn of the twenty-first century, however, because of sharply rising prices for copper—fueled by skyrocketing demand from China and electronics—new milling technologies, and concerns over resource nationalism in developing nations.[7]

Proposed copper-nickel mines faced intense opposition from environmental groups worried about the potential ecological harm of copper-nickel mining and local residents worried that the new mines and their disruption of the landscape would doom northeast Minnesota's growing tourism industry. Although there was more than a century of mining activity on the Iron Range, the proposed new mines would tap into new rock formations. The Duluth Complex's copper-nickel deposits occurred in sulfide-bearing rocks and unlocking the small amounts of valuable copper and nickel would necessarily mean exposing the sulfides as well, creating the potential for pollution from sulfuric acid. The worst problems associated with past copper mining would be averted because the industry no longer used smelters, but environmental groups worried that acids from the mines could pollute water in the Boundary Waters Canoe Area or Lake Superior. As described in chapter 3, the environmental permit process for early taconite mines occurred in the 1940s and 1950s with little organized opposition from environmental groups. By the twenty-first century, however, Minnesota's environmental groups were well funded and well organized and they worked together to oppose the proposed mines. Three of the state's largest environmental groups sponsored a public relations campaign in 2012 to counter what they felt was inaccurate information from the mining industry. Other opponents worried that mining would mar the scenic north-woods landscape and solitude many prized. "This is beautiful country," one local property owner said, "and [the mines] are bound to wreck it." Another said that building copper-nickel mines near the Boundary Waters Canoe Area was "like putting a porn shop next to a church." Other residents opposed the new mines because they threatened their tourism businesses. By 2011, northeast Minnesota's tourism industry was estimated to bring in $700 million in revenue to local businesses, many of which depended on clean water and solitude. Owners of tourism businesses felt they had worked long and hard to create small enterprises that moved the region away from its past resource-extraction economic base and they worried that new mining would destroy their investments. "Our entire lives will have been wasted," one resort owner said about proposed mines. After several decades of trying to move the region away from its dependence on the mining industry, the prospect of new mining encountered opposition from those who imagined a different future for the region.[8]

Local residents supported the proposed mines for equally compelling reasons. Many of those in favor of expanded mining poignantly described families separated when children left the region in search of jobs. New mines raised the possibility, however abstract, that their children might come home again. "I have three grown children living in the [Twin] Cities who I would love to have back in this area," one woman said during a public debate on the new mines, "but that is not possible, we do not have the jobs." Others pointed out that a dependence on tourism had simply replaced dependence on the mining industry. "A viable community has to depend on more than tourism," said one Ely resident.[9]

Although several companies scouted potential mining sites in northeast Minnesota, a venture proposed by PolyMet Corporation attracted the most attention—positive and negative—in the first decades of the twenty-first century. PolyMet is a Canadian mining exploration company focused on developing a copper-nickel mine on the Duluth Complex. The company started exploring the Iron Range for potential mine sites in the 1990s and attracted more attention when it filed for permits to allow mining in 2004—especially on the Iron Range, because it proposed to mine near the former LTV Steel taconite mill that was shut down in 2000. As described earlier, the LTV closure appeared to be yet another tragic shutdown at the time, but the prospect that the old LTV site would spring back to life as a copper-nickel mine heartened many Iron Range residents. PolyMet projected that its mine would employ more than three hundred workers once it was operational. Several hundred more temporary jobs would be created during construction. Many local residents hoped that new industry would step in to replace the jobs lost in the LTV closure and PolyMet's proposed new mine appeared to fulfill that promise. "It would be a return to what, at one point, was normal," said Hoyt Lakes's city administrator, illustrating how the town expected new industry to return. The mine would employ far fewer workers than the 1,400 laid off when LTV closed, but this fact was rarely mentioned in the pro-mining local press. Environmental advocates were concerned, however, that PolyMet would unleash huge quantities of sulphur-bearing tailings and wastewater that could potentially pollute the Lake Superior watershed.[10]

Other proposed mining ventures divided northeast Minnesota communities, including those with long histories of mining. Perhaps no town was more divided than Ely. Founded as a mining camp on the Vermilion Range in the nineteenth century, Ely's mines were gone by the 1960s. The town subsequently developed a substantial trade catering to the many tourists enjoying northwoods vacations in the Boundary Waters Canoe Area. By the early twenty-first century, Ely's main street was lined with outfitters and other businesses catering to the tourist trade. Thus, the prospect of new copper-nickel mining near Ely sharply divided the town between those hoping to return to a mining economy and those committed to expanding the tourist trade. Mining's supporters emphasized the need for high-paying jobs, noting that seasonal jobs in the tourism industry often paid minimum wage. Mining jobs would allow Ely's young people to remain in their hometown, they argued, instead of moving out after high school, as so many young people from the iron mining region had done in the twentieth century. Bartenders in Ely had to ban any discussion of mining because, as one bartender put it, "I don't need to be breaking up fights."[11]

Just as the taconite amendment cleaved the 1960s DFL along urban/rural and ideological/pragmatist lines, recent debate over proposed copper-nickel mines in northeast Minnesota has riven the DFL and opened a wedge for Minnesota's Republicans into the Iron Range. Many Iron Range DFL politicians were increasingly hostile to environmentalists. Outspoken Iron Range politician Tom Rukavina told environmentalists within the DFL Party to "cut the crap and grow up." "I just wish one day that our good DFL senators," he added, "you know, would tell the environmentalists to quit crying wolf, you can't be against everything." Minnesota's DFL could not agree about a party platform supporting new mining projects because of opposition from environmentalists. The party avoided the proposal altogether in 2014, but the issue clearly divided the party. In an echo of strategies used during the 1964 taconite tax debate, Minnesota Republicans used controversy over the proposed PolyMet mine as a wedge issue to divide Iron Range voters from the DFL. Republican politicians adopted the slogans "We're all Iron

Rangers" and "Let's get this done!" to indicate their support for new mining projects and highlight opposition from environmentalists within the DFL Party. The new mining debate deflected attention from the underlying issue: a decline in population and political power on the Iron Range relative to the suburban Twin Cities. Increasingly, the Eighth Congressional District was dominated by the exurbs of St. Paul, not the traditional Iron Range DFL base. A 2014 report noted that DFL politicians were "less worried about the election cycle and more about how to preserve the longtime participation of Iron Rangers in the Democratic-Farmer-Labor Party." The deeper concern was not that the Iron Range would suddenly tip Republican, but rather that the Iron Range's strong support for liberalism would dissolve into dejection and apathy, keeping many out of politics and away from the ballot box altogether. Although Minnesota's Republicans were eager to poach Iron Range votes from the DFL, the underlying reality was that there were simply far fewer voters on the Iron Range by 2014 than there were in the heyday of the Iron Range DFL. Population loss and political apathy threatened to drain the Iron Range of political power altogether, not just shift voters from left to right.[12]

What lessons might history hold for the future of these proposed mines on the Iron Range? It is impossible to accurately predict whether or how these proposed business ventures will come to life, much less how they will or will not affect jobs and the environment in the region. The past century of mining on the Iron Range, however, does hold lessons that both proponents and opponents of the new mining projects would do well to heed.

First, new mining was never the panacea that its supporters claimed, nor did it lead to the utter devastation that critics warned of. As described in chapter 1, Edward W. Davis and his political allies imagined a rich and thriving Iron Range if taconite mining was adopted. A half century after the Iron Range's iron-ore mines switched to taconite, it is abundantly clear that taconite did not solve all of the region's economic woes. A brief construction boom in the late 1960s and early 1970s was quickly undercut by the shutdowns and cutbacks of the 1980s and early 2000s. Despite promises that taconite technology would cleanly and efficiently solve the Iron Range's problems, the truth is that the region's fate was—and will be—determined by the far messier realities of global economics and politics. But it is also important to remember that taconite never fulfilled the apocalyptic rhetoric of some opponents who worried that it would destroy the region's ecosystem. Taconite tailings hardly proved to be the "number one ecological disaster of our time," as Judge Miles Lord worried during the Reserve Mining trial.[13] Just as taconite was neither a cure-all nor a disaster, so too are new mines likely to have a mixed and complicated outcome.

Second, any new mining ventures, especially those involving new minerals, will certainly lead to a host of unexpected and perhaps unwanted consequences. Much of the debate over the environmental consequences of proposed copper-nickel mining in northeast Minnesota has gone back and forth between environmental groups worried about pollution and mining supporters who claim that modern mining can prevent the environmental abuses of the past. Supporters of proposed new mines argue that they will use the latest technology, such as reverse osmosis water treatment or extensive liners, to prevent pollution. Yet it is worth remembering that Reserve Mining Company also proposed to use the latest technology available at the time—directly pumping tailings into Lake Superior based on the prevalent theory that "the secret to pollution is dilution"—to prevent pollution. It is only with hindsight that the error of this approach became clear. Similarly, the problems arising from new copper-nickel mines are unlikely to be easily predicted and thus able to be ameliorated ahead of time. It is the unforeseen consequences that are more worrying, those problems that cannot be known or foreseen precisely because they only appear as the result of complex chains of causality. Hubris that the problems of mining and pollution have been overcome by new technology is almost as old as mining, and just as likely to be proven wrong in the long run.

Conclusion

The Iron Range was just one of many industrial regions in the United States that struggled with deindustrialization and economic decline in the second half of the twentieth century. Taken as a whole, the history of declining industrial regions in the United States during the twentieth century raises a host of provocative questions. As the Iron Range's era of industrial growth fades into the past, how can and should we account for it in the present? Should current policies be aimed at restoring vibrant working-class communities to their central places in American political culture? Or does historical distance instead illustrate the audacity of building communities and political cultures around industry, expecting to create something permanent out of what is necessarily temporary? These questions remain pertinent for the future of the Iron Range and other declining industrial regions in the United States.

The story of the modern Iron Range also raises deeper questions about how decline itself is understood. How can our culture confront long-term economic decline with dignity and a respect for the people making their lives amid decline? In a world that often associates growth with vitality and progress, how can and should we account for the decline and death that are the necessary counterparts of growth? These questions cannot be answered by a single book, but this study points toward new horizons of possibility for thinking about the problem.

INTRODUCTION

1. Robert Shelton, *No Direction Home: The Life and Times of Bob Dylan* (New York: Beech Tree, 1986), 27, 55–56. For more on Dylan's time in Minnesota, see Colleen J. Sheehy and Thomas Swiss, eds., *Highway 61 Revisited: Bob Dylan's Road from Minnesota to the World* (Minneapolis: University of Minnesota Press, 2009); and Toby Thompson, *Positively Main Street: An Unorthodox View of Bob Dylan* (New York: Coward-McCann, 1971; reprint, Minneapolis: University of Minnesota Press, 2008).

2. Edward W. Davis, *Pioneering with Taconite* (St. Paul: Minnesota Historical Society Press, 1964); William T. Hogan, *The 1970s: Critical Years for Steel* (Lexington, Mass.: Lexington Books, 1972). On excitement in the business press, see "Production: A Pellet Gives Iron Ore Industry Shot in the Arm," *Business Week*, December 4, 1965, 107.

3. Jefferson Cowie and Joseph Heathcott, "Introduction: The Meanings of Deindustrialization," in *Beyond the Ruins: The Meanings of Deindustrialization,* ed. Jefferson Cowie and Joseph Heathcott (Ithaca, N.Y.: ILR Press of Cornell University Press, 2003), 4–5.

4. Histories emphasizing the central role of industry in the United States in the twentieth century include Lizabeth Cohen, *Making a New Deal: Industrial Workers in Chicago, 1919–1939* (Cambridge, Mass.: Cambridge University Press, 1990); Michael Denning, *The Cultural Front: The Laboring of American Culture in the Twentieth Century* (London: Verso, 1996); Gary Gerstle, *Working-Class Americanism: The Politics of Labor in a Textile City, 1914–1960* (Princeton, N.J.: Princeton University Press, 2002); and Lary May, *The Big Tomorrow: Hollywood and the Politics of the American Way* (Chicago: University of Chicago Press, 2000).

5. Historians who have emphasized the challenges industrial regions faced in the twentieth century include Jefferson Cowie, *Capital Moves: RCA's Seventy-Year Quest for Cheap Labor* (Ithaca, N.Y.: Cornell University Press, 1999); Thomas Dublin and Walter Licht, *The Face of Decline: Pennsylvania's Anthracite Region in the Twentieth Century* (Ithaca, N.Y.: Cornell University Press, 2005); Steven High, *Industrial Sunset: The Making of North America's Rust Belt, 1969–1984* (Toronto: University of Toronto Press, 2003); and Judith Stein, *Running Steel, Running America: Race, Economic Policy, and the Decline of Liberalism* (Chapel Hill: University of North Carolina Press, 1998).

6. Theodore C. Blegen, *Minnesota: A History of the State,* 2d ed. (Minneapolis: University of Minnesota Press, 1975), 359; David A. Walker, *Iron Frontier: The Discovery and Early Development of Minnesota's Three Ranges* (St. Paul: Minnesota Historical Society Press, 1979), 28–29; Edmund C. Bray, *Billions of Years in Minnesota: The Geological Story of the State* (St. Paul: Science Museum of Minnesota, 1985), 5–6; Minnesota Department of Natural Resources, "Natural History-Minnesota's Geology," www.dnr.state.mn.us/snas/naturalhistory.html.

7. Walker, *Iron Frontier,* 6–7; Agnes M. Larson, *The White Pine Industry in Minnesota: A History* (1949; reprint, Minneapolis: University of Minnesota Press, 2007).

8. Walker, *Iron Frontier,* 16–72. On northeast Minnesota's early history,

see also Marvin Lamppa, *Minnesota's Iron Country: Rich Ore, Rich Lives* (Duluth, Minn.: Lake Superior Port Cities, 2004).

9. Walker, *Iron Frontier*, 73–118.

10. Ibid., 127–230.

11. Arnold R. Alanen, "Years of Change on the Iron Range," in *Minnesota in a Century of Change: The State and Its People since 1900*, ed. Clifford E. Clark Jr. (St. Paul: Minnesota Historical Society Press, 1989), 155–61.

12. Stein, *Running Steel, Running America*, 7–36.

13. On immigration and ethnic identity on the Iron Range, see Alanen, "Years of Change on the Iron Range," 170–80; Clarke A. Chambers, "Welfare on Minnesota's Iron Range," *Upper Midwest History* 3 (1983): 1–40; Mary Lou Nemanic, *One Day for Democracy: Independence Day and the Americanization of Iron Range Immigrants* (Athens: Ohio University Press, 2007); and John Sirjamaki, "The People of the Mesabi Range," in *Selections from "Minnesota History": A Fiftieth Anniversary Anthology*, ed. Rhoda A. Gilman and June Drenning Holmquist (St. Paul: Minnesota Historical Society, 1965), 261–71.

14. Lamppa, *Minnesota's Iron Country*, vii, 245; Aaron Brown, *Overburden: Modern Life on the Iron Range* (Duluth: Red Step, 2008), 4, 14; Alanen, "Years of Change on the Iron Range," 189.

15. There are many similarities between the post–World War II histories of the Iron Range and Appalachia. See Ronald D. Eller, *Uneven Ground: Appalachia since 1945* (Lexington: University Press of Kentucky, 2008); and Alessandro Portelli, *They Say in Harlan County: An Oral History* (New York: Oxford University Press, 2011), 234–365. On Pennsylvania's anthracite mining region, see Dublin and Licht, *The Face of Decline*.

16. Lamppa, *Minnesota's Iron Country*, viii.

17. On the unintended consequences of technology, see Edward Tenner, *Why Things Bite Back: Technology and the Revenge of Unintended Consequences* (New York: Knopf, 1996).

18. Other local and regional case studies of declining industrial regions include Eric L. Clements, *After the Boom in Tombstone and Jerome, Arizona: Decline in Western Resource Towns* (Reno: University of Nevada Press, 2003); Dublin and Licht, *The Face of Decline*; David Koistinen, *Confronting Decline: The Political Economy of Deindustrialization in Twentieth-Century New England* (Gainesville: University Press of Florida, 2013); and Guian A. McKee, *The Problem of Jobs: Liberalism, Race, and Deindustrialization in Philadelphia* (Chicago: University of Chicago Press, 2008). One notable study that does address the relationship between technological change and employment in postwar industry is David F. Noble, *Forces of Production: A Social History of Industrial Automation* (New York: Oxford University Press, 1986), although Noble focuses on efforts to control workers rather than on technological efforts to save industry.

19. On taxation's role in policy, see Robin L. Einhorn, *American Taxation, American Slavery* (Chicago: University of Chicago Press, 2006); and Peter A. Shulman, "The Making of a Tax Break: The Oil Depletion Allowance, Scientific Taxation, and Natural Resources Policy in the Early Twentieth Century," *Journal of Policy History* 23, no. 3 (2011): 281–322.

20. On postwar economic development policy in the United States, see Otis L. Graham, *Losing Time: The Industrial Policy Debate* (Cambridge: Harvard University Press, 1992); Gregory S. Wilson, *Communities Left Behind: The Area Redevelopment Administration, 1945–1965* (Knoxville: University of Tennessee Press, 2009); and Andrew L. Yarrow, *Measuring America: How Economic Growth Came to Define American Greatness in the Late Twentieth Century* (Amherst: University of Massachusetts Press, 2010).

21. On the labor–liberal alliance in the postwar United States, see Andrew Battista, *The Revival of Labor Liberalism* (Urbana: University of Illinois Press, 2008); Kevin Boyle, *The UAW and the Heyday of American Liberalism, 1945–1968* (Ithaca, N.Y.: Cornell University Press, 1995); and Nelson Lichtenstein, *The Most Dangerous Man in Detroit: Walter Reuther and the Fate of American Labor* (New York: Basic Books, 1995).

22. David W. Blight, *Race and Reunion: The Civil War in American Memory* (Cambridge: Belknap Press of Harvard University Press, 2001); Cathy Stanton, *The Lowell Experiment: Public History in a Postindustrial City* (Amherst: University of Massachusetts Press, 2006).

23. Dan Kraker, "On the Iron Range, Debating Whether Long-Term Prosperity Follows More Mining," Minnesota Public Radio News, March 4, 2013: http://minnesota.publicradio.org/display/web/2013/03/02/regional/minnesota-mining-jobs (accessed March 6, 2013).

24. Norwood B. Melcher and John Hozik, "Iron Ore," in *United States Bureau of Mines Minerals Yearbook 1945*, ed. E. W. Pehrson and H. D. Keiser (Washington, D.C.: Government Printing Office, 1947), 566; Christopher A. Tuck and Robert L. Virta, "Iron Ore," in *U.S. Geological Survey Minerals Yearbook 2011* (Washington, D.C.: U.S. Geological Survey, 2013), table 2.

25. Historian Steven High has similarly criticized the use of boom-and-bust rhetoric for explaining economic change in Canada's peripheral regions (Steven High, "'The Wounds of Class': A Historiographical Reflection on the Study of Deindustrialization, 1973–2012," *History Compass* 11, no. 11 [2013]: 1002).

26. On the timing of deindustrialization and arguments that it began well before the 1970s and 1980s, see Cowie, *Capital Moves*; Dublin and Licht, *The Face of Decline*; Koistinen, *Confronting Decline*; and Thomas J. Sugrue, *The Origins of the Urban Crisis: Race and Inequality in Postwar Detroit* (Princeton, N.J.: Princeton University Press, 1996).

27. The best overview of scholarship on deindustrialization is High, "The Wounds of Class," and Cowie and Heathcott, "Introduction." For an example of one community that did successfully fight to preserve local industry, see Perry Bush, *Rust Belt Resistance: How a Small Community Took on Big Oil and Won* (Kent, Ohio: Kent State University Press, 2012).

28. Christopher H. Johnson, *The Life and Death of Industrial Languedoc, 1700–1920* (New York: Oxford University Press, 1995); Tim Strangleman, James Rhodes, and Sherry Linkon, "Introduction to Crumbling Cultures: Deindustrialization, Class, and Memory," *International Labor and Working-Class History* 84 (2013): 12; Dublin and Licht, *The Face of Decline*. There is a large literature on deindustrialization outside the United States. Examples from Canada include

High, *Industrial Sunset*; and Ian McKay, *The Quest of the Folk: Antimodernism and Cultural Selection in Twentieth-Century Nova Scotia* (Montreal: McGill-Queen's University Press, 1994). For studies on Europe, see Jackie Clarke, "Closing Moulinex: Thoughts on the Visibility and Invisibility of Industrial Labour in Contemporary France," *Modern and Contemporary France* 19, no. 4 (2011): 443–58; John Kirk, Sylvie Contrepois, and Steve Jefferys, eds., *Changing Work and Community Identities in European Regions: Perspectives on the Past and Present* (New York: Palgrave Macmillan, 2012); and Tim Strangleman, *Work Identity at the End of the Line?: Privatisation and Culture Change in the UK Rail Industry* (New York: Palgrave Macmillan, 2004).

29. Several recent studies of deindustrialization emphasize the role of the state in fighting industrial decline. See Joseph Heathcott and Máire Agnes Murphy, "Corridors of Flight, Zones of Renewal: Industry, Planning, and Policy in the Making of Metropolitan St. Louis, 1940–1980," *Journal of Urban History* 31, no. 2 (2005): 151–89; Koistinen, *Confronting Decline*; McKee, *The Problem of Jobs*; and Judith Stein, *Pivotal Decade: How the United States Traded Factories for Finance in the Seventies* (New Haven: Yale University Press, 2010). In the 1980s, there was a brief but vigorous debate over industrial policy that also addressed the federal government's role in fighting deindustrialization. See Graham, *Losing Time*.

1. A TECHNOLOGICAL FIX?

1. Davis, *Pioneering with Taconite*, 183.

2. On the technological fix, see Lisa Rosner, ed., *The Technological Fix: How People Use Technology to Create and Solve Problems* (New York: Routledge, 2004), especially Timothy LeCain's chapter.

3. On technological and scientific optimism in the twentieth century, see Thomas J. Misa, *Leonardo to the Internet: Technology and Culture from the Renaissance to the Present*, 2d ed. (Baltimore: Johns Hopkins University Press, 2011), 128–89.

4. Logan Hovis and Jeremy Mouat, "Miners, Engineers, and the Transformation of Work in the Western Mining Industry, 1880–1930," *Technology and Culture* 37, no. 3 (1996): 434–35; Jeffrey T. Manuel, "Mr. Taconite: Edward W. Davis and the Promotion of Low-Grade Iron Ore, 1913–1955," *Technology and Culture* 54, no. 2 (April 2013): 317–45. For other perspectives on this seismic shift in the mining industry, see Roger Burt, "Innovation or Imitation? Technological Dependency in the American Nonferrous Mining Industry," *Technology and Culture* 41, no. 2 (2000): 321–47; Timothy J. LeCain, *Mass Destruction: The Men and Giant Mines that Wired America and Scarred the Planet* (New Brunswick, N.J.: Rutgers University Press, 2009), 118–19; and A. B. Parsons, ed., *Seventy-five Years of Progress in the Minerals Industry, 1871–1946* (New York: American Institute of Mining and Metallurgical Engineers, 1947).

5. Robert H. Richards, *A Text Book of Ore Dressing* (New York: McGraw-Hill, 1909), 393; Edwin C. Eckel, *Iron Ores: Their Occurrence, Valuation, and Control* (New York: McGraw-Hill, 1914), 157–60, 251–63. Edward Davis later attributed the rare use of magnetic separation to the opening of the rich Mesabi Range at the turn of the century. As Davis put it, "after the Mesabi Range

started large scale production in 1900, all attempts at magnetic roasting and concentration seem to have ceased. The process was, at best, expensive and apparently could not compete with the raw ores" (Edward W. Davis, *Magnetic Concentration of Iron Ore,* Bulletin of the University of Minnesota School of Mines Experiment Station 9 [Minneapolis: University of Minnesota, 1921], 38).

6. Matthew Josephson, *Edison: A Biography* (New York: McGraw-Hill, 1959), 379; Neil Baldwin, *Edison: Inventing the Century* (New York: Hyperion, 1995), 215; Paul Israel, *Edison: A Life of Invention* (New York: John Wiley and Sons, 1998), 338–39; and Jill Jonnes, *Empires of Light: Edison, Tesla, Westinghouse, and the Race to Electrify the World* (New York: Random House, 2003), 349. See also George S. May and Victor F. Lemmer, "Thomas Edison's Experimental Work with Michigan Iron Ore," *Michigan History* 53, no. 2 (1969): 111–13. The best scholarly treatment of Edison's iron-ore venture is W. Bernard Carlson, "Edison in the Mountains: The Magnetic Ore Separation Venture, 1879–1900," *History of Technology* 8 (1983): 37–59.

7. On flotation, see Hovis and Mouat, "Miners, Engineers, and the Transformation of Work in the Western Mining Industry, 1880–1930"; Manuel, "Mr. Taconite"; Jeremy Mouat, "The Development of the Flotation Process: Technological Change and the Genesis of Modern Mining, 1898–1911," *Australian Economic History Review* 36, no. 1 (1996): 3–31; Otis Young and Robert Lenon, *Western Mining: An Informal Account of Precious-Metals Prospecting, Placering, Lode Mining, and Milling on the American Frontier from Spanish Times to 1893* (Norman: University of Oklahoma Press, 1970), 231–33; Arthur F. Taggart, "Seventy-five Years of Progress in Ore Dressing," in Parsons, *Seventy-five Years of Progress in the Mineral Industry, 1871–1946,* 101–6; A. B. Parsons, *The Porphyry Coppers* (New York: American Institute of Mining and Metallurgical Engineers, 1933), 441–63; and LeCain, *Mass Destruction,* 165–68.

8. Davis, *Pioneering with Taconite,* 18, 20–23. For an overview of Davis's early experimental work, see Davis, *Magnetic Concentration of Iron Ore.* See also "Magnetic Concentration Apparatus Developed at the Experiment Station," undated manuscript, Mines Experiment Station-Magnetic Concentration Folder, Box 11, Mines Experiment Station records, University Archives, University of Minnesota, Minneapolis, Minnesota (hereafter MES Records); and James Baker Ross, "Taconite: The Science of Design," PhD dissertation, University of Minnesota, 1989.

9. Davis, *Pioneering with Taconite,* 20. In later correspondence with historian Victor Lemmer, Davis claimed that he was "somewhat familiar with Edison's work" on iron-ore separation but insisted that his methods could not have been used for taconite separation. See May and Lemmer, "Thomas Edison's Experimental Work with Michigan Iron Ore," 130n53.

10. Davis, *Pioneering With Taconite,* 67; Ross, "Taconite," 133–34. On mass-removal copper mining, see LeCain, *Mass Destruction*; Parsons, *The Porphyry Coppers*; and Charles K. Hyde, *Copper for America: The United States Copper Industry from Colonial Times to the 1990s* (Tucson: University of Arizona Press, 1998). For the earlier history of copper mining in the Lake Superior region, see Larry Lankton, *Cradle to Grave: Life, Work, and Death at the Lake Superior*

Copper Mines (New York: Oxford University Press, 1991); Larry Lankton, *Beyond the Boundaries: Life and Landscape at the Lake Superior Copper Mines, 1840–1875* (New York: Oxford University Press, 1997); and Larry D. Lankton, *Hollowed Ground: Copper Mining and Community Building on Lake Superior, 1840s–1990s* (Detroit: Wayne State University Press, 2010). For the major developments in comminution during these years, see Taggart, "Seventy-five Years of Progress in Ore Dressing," 90–95.

11. Davis, *Pioneering with Taconite*, 22–23, 38; and Davis, *Magnetic Concentration of Iron Ore*, 12–85. Despite Davis's borrowing from copper milling, he remained committed to magnetic separation rather than turning to flotation. This decision was questioned by some of his fellow mining engineers and Davis exchanged letters with them during the 1920s to defend his use of magnetic separation. See, for example, F. H. Wilcox letter to E. W. Davis, December 30, 1922, Freyn, Brassert & Company Folder, Box 2, MES Records. Other engineers did develop successful flotation processes for iron ores, although they were often more expensive than magnetic separation (Davis, *Pioneering with Taconite*, 109, 192; Terry S. Reynolds and Virginia P. Dawson, *Iron Will: Cleveland-Cliffs and the Mining of Iron Ore, 1847–2006* [Detroit: Wayne State University Press, 2011], 165, 175–77).

12. On Jackling, see LeCain, *Mass Destruction*; and Parsons, *The Porphyry Coppers*.

13. W. G. Swart letter to D. C. Jackling, April 7, 1918, in Mesabi Iron Company, New York, Misc. Papers, 1917–1919 Folder, Mesabi Iron Company records, Minnesota Historical Society, St. Paul, Minnesota (hereafter MIC records); Hayden, Stone, and Co., "Mesabi Iron Company," undated letter; in Mesabi Iron Company, New York, Misc. Papers, 1921–1923 Folder, MIC Records; Hyde, *Copper for America*, 141; Parsons, *The Porphyry Coppers*, 76; LeCain, *Mass Destruction*, 6.

14. Hyde, *Copper for America*, 139–45; Davis, *Pioneering with Taconite*, 26, 28, 42, 56. Swart later prepared several hundred pages of reports on the firm's mining and milling processes, including detailed evaluations of the technical success and costs of each machine. See W. G. Swart Reports Folder, MIC Records.

15. Davis, *Pioneering with Taconite*, 53–55. On the transition from the Bessemer process to open-hearth steelmaking technology in the United States, see Thomas J. Misa, *A Nation of Steel: The Making of Modern America, 1865–1925* (Baltimore: Johns Hopkins University Press, 1995), 56–59, 76–84, 152–53.

16. D. C. Jackling letter to Mesabi Iron Company stockholders, May 8, 1924, in Miscellaneous Papers, 1924–1953 Folder, MIC Records; Davis, *Pioneering with Taconite*, 59.

17. Davis, *Pioneering with Taconite*, 65.

18. Peter Temin, *Iron and Steel in Nineteenth-Century America: An Economic Inquiry* (Cambridge: MIT Press, 1964), 197; Paul H. Landis, *Three Iron Mining Towns: A Study in Cultural Change* (1938; reprint, New York: Arno Press, 1970), 107; E. D. Gardner and McHenry Mosier, *Open-Cut Metal Mining* (Washington, D.C.: Government Printing Office, 1941), 70–71.

19. Duane A. Smith, *Mining America: The Industry and the Environment, 1800–1980* (Lawrence: University Press of Kansas, 1987), 86, 103.

20. LeCain, *Mass Destruction,* 159; C. W. Hayes, *Iron Ores of the United States* (Washington, D.C.: United States Geological Survey, 1909), 3:520.

21. David E. Nye, *American Technological Sublime* (Cambridge: MIT Press, 1994), 126.

22. William Stanley Jevons, *The Coal Question: An Inquiry concerning the Progress of the Nation, and the Probable Exhaustion of Our Coal-Mines* (London: Macmillan, 1865).

23. Hayes, *Iron Ores of the United States,* 520, 490; Edwin C. Eckel, *Iron Ores: Their Occurrence, Valuation, and Control* (New York: McGraw-Hill, 1914), 202.

24. LeCain, *Mass Destruction,* 124.

25. Edward W. Davis, *The Future of the Lake Superior District as an Iron-Ore Producer,* Bulletin of the University of Minnesota School of Mines Experiment Station 7 (Minneapolis: University of Minnesota, 1920), 3.

26. Ibid., 5.

27. Edward W. Davis, "The Iron Mining Industry in Minnesota," transcript of speech to Six O'Clock Club, Minneapolis, January 14, 1924, in "Articles by E. W. Davis and M. E. S. Staff" Folder, Box 13, MES Records.

28. E. W. Davis, "Progress with Taconite," transcript of speech to Hibbing Chamber of Commerce, February 21, 1950, 2; Box 13, MES Records; E. W. Davis letter to Fred Cina, April 24, 1950; Box 1, Folder 02.03.01.00000010; Fred A. Cina Papers Collection, Iron Range Research Center, Chisholm, Minnesota (hereafter Cina Papers).

29. Edward W. Davis, *Lake Superior Iron Ore and the War Emergency: A Report Presented to the Materials Division of the War Production Board* (Minneapolis: University of Minnesota Mines Experiment Station, 1942), 6; Davis, *Pioneering with Taconite,* 110–111; E. W. Davis to Howard Young, April 10, 1943, Reserve Mining Project General Folder, Box 10, MES Records. See also Alfred E. Eckes Jr., *The United States and the Global Struggle for Minerals* (Austin: University of Texas Press, 1979), 121–98.

30. Davis, *Lake Superior Iron Ore and the War Emergency,* 6; Davis, *Pioneering with Taconite,* 110–11.

31. University of Minnesota Mineral Resources Research Station, *History of the Mineral Resources Research Center* (Minneapolis, 1981), 64; Edward W. Davis, "A Problem in Taxation," transcript of speech at untitled location, 1923, 4, in "Articles by E. W. Davis and M.E.S. Staff" Folder, Box 13, MES Records; Davis, *Pioneering with Taconite,* 93–94.

32. Charles L. Horn to Edward W. Davis, July 2, 1947, Minneapolis, Minnesota, General Folder, Box 11, MES Records. See also Horn's earlier letter, Charles L. Horn to Edward W. Davis, June 23, 1947; Benjamin F. Fairless, "Iron Ore Is Available," transcript of remarks to Subcommittee on the Study of Monopoly Power of the House Committee on the Judiciary, Washington, D.C., April 26–28, 1950, in Reconstruction Finance Corp. Act Folder, Box 15, MES Records.

33. Davis, *Pioneering with Taconite,* 92–94, 98.

34. Ibid., 99–103.

35. Ibid. 100, 104. See also the transcript of E. W. Davis, *Edward W. Davis v. Nordahl I. Onstad,* Minneapolis, March 8–10, 1960, 26; Edward Wilson Davis papers, University Archives, University of Minnesota, Minneapolis, Minnesota (hereafter Davis Papers; note that these are different from the Edward W. Davis papers—abbreviated as EWD Papers—held at the Minnesota Historical Society). For the text of the revised statute, see An Act Relating to the Taxation of Taconite (April 22, 1941), *Session Laws of Minnesota for 1941,* chapter 375, https://www.revisor.mn.gov/laws/.

36. Ross, "Taconite," 229–30; "This Taconite Kit Is Sent to You by the Mines Experiment Station of the University of Minnesota," March 21, 1951, University of Minnesota School of Mines "Misc." Folder, Box 10, MES Records; Saul L. Wernick and David H. Gaines, transcript of "University of Minnesota Television Program: Taconite," WTGN-TV, February 15, 1951, 14, Mines Experiment Station "Misc." Folder, Box 10, MES Records.

37. Orville L. Freeman to E. W. Davis, November 3, 1958, Correspondence 1911–1973, Folder 2, Box 1, EWD Papers.

38. Edward W. Davis to K. C. McCutcheon, June 23, 1948, Kentucky Armco Steel Corp. Folder, Box 10, MES Records.

39. Edward W. Davis, "Fear of Tax Increase Declared Holding Back Taconite Industry," *Minneapolis Star,* October 30, 1958. On corporate taxation during this era, see Cathie J. Martin, *Shifting the Burden: The Struggle over Growth and Corporate Taxation* (Chicago: University of Chicago Press, 1991).

40. Edward W. Davis, "Mesabi Ore Situation," transcript of speech, Eveleth, Minnesota, April 5, 1946, 1, Box 13, MES Records; Edward W. Davis, "Taconite," transcript of radio broadcast, WEBC, Duluth, Minnesota, January 30, 1948, 3, Box 13, MES Records; Edward W. Davis, "Eveleth Meeting" notes, 1957, Background Material, 1912–1968 Folder, Box 1, EWD Papers.

41. Donald T. Harries to R. T. Elstad, June 4, 1954; in Taconite 1954 Folder, Box 1987.0970.MSS.00000011, Butler Brothers and Hanna Mining Company Records Collection, Iron Range Research Center, Chisholm, Minnesota (hereafter BBHMC Records).

42. Davis, *Pioneering with Taconite,* 118, 142–43; "$75,000,000 Plant for Taconite Set," *New York Times,* September 22, 1951; "Erie Mining Plans $300,000,000 Plant," *New York Times,* February 17, 1952; Thomas E. Mullaney, "Big Boom Foreseen in Low-Grade Ores," *New York Times,* October 14, 1949; Reynolds and Dawson, *Iron Will,* 169–204.

43. Davis, *Pioneering with Taconite,* 145, 176–77, 183; Thomas R. Huffman, "Exploring the Legacy of Reserve Mining: What Does the Longest Environmental Trial in History Tell Us about the Meaning of American Environmentalism?" *Journal of Policy History* 12, no. 3 (2000): 340; Dwight Eisenhower to Reserve Mining, September 12, 1956, Correspondence 1911–1973 Folder 2, Box 1, EWD Papers.

44. Davis, *Pioneering with Taconite,* 157–58, 168; Reserve Mining Company, "Your Visit to Reserve," undated pamphlet, in Edward W. Davis Info Folder, held at the University of Minnesota's University Archives.

45. Davis, *Pioneering with Taconite,* 181; Huffman, "Exploring the Legacy of Reserve Mining," 340; Hogan, *The 1970s,* 83.

46. "Davis Earns Tribute," *Range Facts* (Virginia, Minnesota), August 22, 1957; D. J. Tice, "The Thing on the Hill," *Corporate Report Minnesota*, October 1982, 62–63; "Davis Given Steel Award," *Silver Bay News*, May 5, 1971.

47. "Taconite Setback," *Minneapolis Star*, February 26, 1958; "Reserve Mining to Cut Output of Taconite Pellets," *Minneapolis Tribune*, February 25, 1958.

48. On the effects of pelletized magnetic ore on other mining regions, especially those with nonmagnetic ore that was more expensive to pelletize, see Peter J. Kakela, "The Shift to Taconite Pellets: Necessary Evil or Lucky Break?" *Michigan History Magazine* (November/December 1994), 70–75; Peter J. Kakela, Allen K. Montgomery, and William C. Patric, "Factors Influencing Mine Location: An Iron Ore Example," *Land Economics* 58, no. 4 (1982): 524–36; and Peter Frank Mason, "Some Changes in Domestic Iron Mining as a Result of Pelletization," *Annals of the Association of American Geographers* 59, no. 3 (1969): 535–51. See also Davis, *Pioneering with Taconite*, 194–95.

2. CREATING A TAX-CUT CONSENSUS ON THE IRON RANGE

1. On U.S. tax policy in the twentieth century, see W. Elliot Brownlee, *Federal Taxation in America: A Short History* (New York: Woodrow Wilson Center and Cambridge University Press, 1996). Brownlee focuses exclusively on federal taxes rather than the emphasis on state taxes in this chapter. See also the useful appendix in Robin L. Einhorn, *American Taxation, American Slavery* (Chicago: University of Chicago Press, 2006), 257–67.

2. Joseph A. Schumpeter, "The Crisis of the Tax State" (1954), quoted in Einhorn, *American Taxation, American Slavery*, 6; Alice Kessler-Harris, *In Pursuit of Equity: Women, Men, and the Quest for Economic Citizenship in 20th-Century America* (New York: Oxford University Press, 2001), 170–202; Julian E. Zelizer, *Taxing America: Wilbur D. Mills, Congress, and the State, 1945–1975* (New York: Cambridge University Press, 1998); Brownlee, *Federal Taxation in America*, 4.

3. Julian Zelizer, "The Uneasy Relationship: Democracy, Taxation, and State Building since the New Deal," in *The Democratic Experiment: New Directions in American Political History*, ed. Meg Jacobs, William J. Novak, and Julian Zelizer (Princeton, N.J.: Princeton University Press, 2003), 288; Cathie J. Martin, *Shifting the Burden: The Struggle over Growth and Corporate Taxation* (Chicago: University of Chicago Press, 1991), 10.

4. On tax protests in the nineteenth century, see Sven Beckert, "Democracy in the Age of Capital: Contesting Suffrage Rights in Gilded Age New York," in Jacobs, Novak, and Zelizer, *The Democratic Experiment*, 146; Heather Cox Richardson, *West from Appomattox: The Reconstruction of America after the Civil War* (New Haven: Yale University Press, 2007). On the post–World War II tax revolt, see Robert Self, *American Babylon: Race and the Struggle for Postwar Oakland* (Princeton, N.J.: Princeton University Press, 2005).

5. Donald G. Sofchalk, "The Iron Miners' Struggle for Collective Bargaining, 1941–1943," *Labor's Heritage* 5, no. 1 (1993): 69. On the broader effort to organize the steel industry in the 1930s, see Robert H. Zieger, *American Workers, American Unions*, 2d ed. (Baltimore: Johns Hopkins University Press,

1994), 45–51. On the 1907 and 1916 strikes, see Arnold R. Alanen, "Early Labor Strife on Minnesota's Mining Frontier," *Minnesota History* 52 (1991): 247–53; Alanen, "Years of Change on the Iron Range," 181–85; Neil Betten, "Strike on the Mesabi-1907," *Minnesota History* 40 (1967): 340–47; Neil Betten, "Riot, Revolution, and Repression in the Iron Range's Strike of 1916," *Minnesota History* 41 (1968): 82–94; Robert Eleff, "The 1916 Minnesota Miners' Strike against U.S. Steel," *Minnesota History* 51 (1988): 63–74; and Donald G. Sofchalk, "Organized Labor and the Iron Ore Miners of Northern Minnesota, 1907–1936," *Labor History* 12 (1971): 225–42.

6. Sofchalk, "The Iron Miners' Struggle for Collective Bargaining, 1941–1943," 69–75. On industrial unionism during the early postwar decades, see Zieger, *American Workers, American Unions*, 137–67.

7. W. Elliot Brownlee, "Tax Regimes, National Crisis, and State-Building in America," in *Funding the Modern American State, 1941–1995: The Rise and Fall of the Era of Easy Finance*, ed. W. Elliot Brownlee (Cambridge: Cambridge University Press, 1996), 93; Kessler-Harris, *In Pursuit of Equity*, 172–73.

8. Meg Jacobs, *Pocketbook Politics: Economic Citizenship in Twentieth-Century America* (Princeton, N.J.: Princeton University Press, 2005); Zelizer, *Taxing America*.

9. James Reston, "Mine-Job Security Seen in 'Stockpile,'" *New York Times*, July 16, 1946. On Minnesota's state politics, see John E. Haynes, "Reformers, Radicals, and Conservatives," in Clark, *Minnesota in a Century of Change*, 361–96.

10. "Mining: Tension on the Mesabi," *Newsweek*, May 1, 1961, 70. On the culture war narrative, see Thomas Frank, *What's the Matter with Kansas? How Conservatives Won the Heart of America* (New York: Metropolitan Books, 2004); Rick Perlstein, *Nixonland: The Rise of a President and the Fracturing of America* (New York: Scribner, 2008).

11. Roy G. Blakey, *Taxation in Minnesota* (Minneapolis: University of Minnesota Press, 1932), 234–76, 600; Walker, *Iron Frontier*, 37, 131; Lewis E. Young, *Mine Taxation in the United States* (Urbana: University of Illinois Press, 1917), 54–55.

12. Blakey, *Taxation in Minnesota*, 247; Brownlee, *Federal Taxation in America*, 5.

13. Warren Roberts, *State Taxation of Metallic Deposits* (Cambridge: Harvard University Press, 1944), 342–53.

14. Alanen, "Years of Change on the Iron Range," 165.

15. Landis, *Three Iron Mining Towns*, 112; Dana H. Miller, *The Iron Range Resources and Rehabilitation Board: The First Fifty Years* (Eveleth, Minn.: IRRRB, 1991), 3–4; Lorena Hickok, *One Third of a Nation: Lorena Hickok Reports on the Great Depression*, ed. Richard Lowitt and Maurine H. Beasley (Urbana: University of Illinois Press, 2000), 129–30.

16. Nathan Cohen, "Razzle-Dazzle Village," *American Mercury*, March 1944, 346–50.

17. George L. Peterson, "Hibbing Disclaims the Title of 'Razzle Dazzle Village,'" *Minneapolis Star*, May 22, 1944; Andrew Bradish letter to Fred Cina, February 28, 1949, Box 1, Folder 02.03.01.00000010, Cina Papers.

18. Davis, *Pioneering with Taconite*, 189–90.

19. "Constitutional Balm for an Old Sore," *Business Week*, September 14, 1963, 112; "Lake Superior Iron Posts Market Loss," *New York Times*, August 5, 1960; "Mining: The Iron Amendment," *Newsweek*, November 16, 1964, 74.

20. Leo J. Hertzel, "Defeat on the Great Lakes," *Nation*, September 8, 1962, 109–11.

21. "U.S. Iron Ore Market Squeezed by Steel Slump, Rise in Imports," *New York Times*, March 23, 1958; "Mesabi Iron Work Cut," *New York Times*, March 13, 1958; "U.S. Steel Closes Several Furnaces as Demand Slides," *New York Times*, March 23, 1960. On mining employment numbers in Minnesota, see United States Geological Survey, "Iron Ore Statistical Compendium," table 6, http://minerals.usgs.gov/minerals/pubs/commodity/iron_ore/stat/tbl6.txt for the years from 1965 to 1990. Mining employment numbers prior to 1965 are available from the yearly *Minerals Yearbook* compiled by the U.S. Bureau of Mines, and are available online through the University of Wisconsin Digital Collections Ecology and Natural Resources Collection, http://uwdc.library.wisc.edu/collections/EcoNatRes/MineralsYearBk.

22. Clay Blair Jr., "Minnesota Grows Older," *Saturday Evening Post*, March 18, 1961; "Iron Mines Face a Bleak Outlook," *New York Times*, April 10, 1961.

23. Homer Bigart, "Poverty Blights Iron Ore Region," *New York Times*, January 13, 1964. Population statistics are from Historical Minnesota County, City, and Township Population Data, Center for Small Towns, University of Minnesota Morris, http://www.morris.umn.edu/services/cst/dar/mdc/census/index.php; "4 Iron Ore Towns Told to Combine," *New York Times*, February 9, 1964.

24. G. A. Kaltenbach letter to "Christian Friends," October 5, 1960, in Misc. 1960 Folder, Box 1987.0970.MSS.00000022, BBHMC Records.

25. Bob Dylan, "North Country Blues," *The Times They Are A-Changin'* (Compact disc: Columbia, 1964).

26. See transcript "In the Matter of the Equalization of Assessments of Mined and Unmined Iron Ore for the Year 1957," October 25, 1957, Box 14, Folder 530, Cina Papers.

27. "American Capital in Canada—and a Boom in Iron Ore," *U.S. News and World Report*, May 4, 1964, 72, 74; Herbert L. Matthews, "Knob Lake Iron Ore Will Help Lift Canada to Third Largest Producer," *New York Times*, August 8, 1953.

28. "Ore Coming Soon from Venezuela," *New York Times*, August 29, 1949; "Big Iron Ore Body Seen in Venezuela," *New York Times*, January 29, 1950. On the development of African iron-ore mines, see "A Jungle Mesabi," *Fortune* (December 1961): 81–82, 84; "Africa: A Mountain of Riches," *Time*, October 25, 1963, 99; John M. Lee, "Big Finds of Ore Add to Reserves," *New York Times*, November 24, 1963.

29. Lee, "Big Finds of Ore Add to Reserves." On postwar international capital flight, see Cowie, *Capital Moves*; High, *Industrial Sunset*; and Stein, *Pivotal Decade*.

30. "Brazil: Hanna's Immovable Mountains," *Fortune* (April 1965): 55–56, 63.

31. Charles F. Park Jr., "Iron and Steel," speech reprinted in *The Twenty-seventh, Twenty-eighth and Twenty-ninth Warren Lectures on Sources of Energy, A Mineral Policy, and Iron and Steel* (Minneapolis: University of Minnesota School of Mines, 1959), 8.

32. IRRRB, *Biennial Report of the Iron Range Resources and Rehabilitation Commission, 1943–1945* (St. Paul: IRRRB, 1946), 21.

33. E. W. Davis, "Fear of Tax Increase Declared Holding Back Taconite Industry," *Minneapolis Star,* October 30, 1958.

34. E. W. Davis, "Eveleth Meeting" notes, 1957. Background Material, 1912–1968 Folder, Box 1, EWD Papers.

35. "Constitutional Taconite Tax Limit Asked," *Minneapolis Tribune,* December 3, 1960; Sam Newlund, "300 Pack House Hearing, Back Taconite Tax Relief," *Minneapolis Tribune,* May 9, 1961.

36. Leonard Inskip, "State Ore Tax Policy Criticized," *Minneapolis Tribune,* September 25, 1960.

37. "Mining: Tension on the Mesabi," *Newsweek,* May 1, 1961, 70, 75; "Iron Imports Vex Mining States," *Business Week,* August 9, 1958, 109.

38. "Steel: Taconite Boom," *Time,* April 28, 1952, 92–94. The specific financing program was a $298-million "quick tax write-off" by the Defense Production Administration, the largest such write-off the administration had granted up to that point. On the Defense Production Act, see Eckes, *The United States and the Global Struggle for Minerals,* 147–73; Sam Newlund, "Iron Officials: Taconite Tax Change Vital," *Minneapolis Tribune,* May 10, 1961.

39. Remarks of Senator John F. Kennedy, Hibbing, Minnesota, October 2, 1960, John F. Kennedy Presidential Library and Museum, http://www.jfklibrary.org/Research/Research-Aids/JFK-Speeches/Hibbing-MN_19601002.aspx.

40. "Cina Turns GOP Iron Ore Issue into Boomerang," *Labor World,* August 7, 1958.

41. "DFL Opposes Amendment to Aid Ore Firms," *Minneapolis Tribune,* April 20, 1961.

42. "GOP, DFL Clash Anew on Taconite," *Minneapolis Tribune,* October 21, 1961; John C. McDonald, "Fraser Calls Taconite Law Easiest," *Minneapolis Tribune,* October 22, 1961; John C. McDonald, "Big Range Issue Is Taconite Plan," *Minneapolis Tribune,* September 20, 1961.

43. John C. McDonald, "Taconite Proposal Up Again," *Minneapolis Tribune,* April 30, 1961.

44. Donald D. Wozniak, transcript of speech to USWA local 1663, Hibbing, Minnesota, November 26, 1963, in Taconite Amendment 1961 Folder, Box 1987.0970.MSS.00000127, BBHMC Records.

45. KDAL, "In Depth," transcript of radio broadcast, undated, in Taconite Amendment publicity Folder, Box 2, EWD Papers.

46. George Farr, transcript of testimony to Committee on Taxes, May 11, 1961, Box 1, Folder 02.07.02.00000028, Cina Papers; Sam Newlund, "Vote Kills Taconite Tax Plan," *Minneapolis Tribune,* May 12, 1961; "No Taconite Guarantee" editorial, *Minneapolis Tribune,* May 14, 1961; "Taconite Amendment

Drive Pleases Andersen," *Minneapolis Tribune*, August 13, 1961; "DFL States Reason for Taconite Stand," *Minneapolis Tribune*, October 23, 1961.

47. "Mining: Tension on the Mesabi," *Newsweek*, May 1, 1961, 70; Gilbert Finnegan letter to Fred Cina, April 27, 1961, Box 1, Folder 02.03.01.00000012, Cina Papers; "Equalization Bill Draws Fire on Iron Range," *Minneapolis Tribune*, March 27, 1961.

48. Fred Cina letter to Harry Zinsmaster, February 8, 1961, Box 1, Folder 02.03.01.00000012, Cina Papers.

49. Sam Romer, "AFL–CIO Opposes Proposed 'Freeze' on Taconite Taxes," *Minneapolis Tribune*, December 6, 1960; Sam Newlund, "Senate Tax Unit Supports Taconite Equity Guarantee," *Minneapolis Tribune*, April 12, 1961; Sam Romer, "Drive Set on Taconite Proposal," *Minneapolis Tribune*, September 17, 1961.

50. "Majority Would Tax Taconite like Other Industries in State," *Minneapolis Tribune*, May 14, 1961; Mercer Cross, "Andersen Calls for Taconite Aid Action," *Minneapolis Tribune*, August 4, 1961; petition to Fred Cina, March 13, 1961, Box 1, Folder 02.03.01.00000012, Cina Papers. This folder contains many letters and telegrams urging Cina to support the amendment.

51. Sam Newlund, "300 Pack House Hearing, Back Taconite Tax Relief," *Minneapolis Tribune*, May 9, 1961.

52. Transcript of testimony to Minnesota House Committee on Taxes, regarding HF 116, Taconite Constitutional Amendment, St. Paul, May 8–9, 1961, 10, in Taxes Equalization Program 1961 Folder, Box 1987.0970.MSS.00000024, BBHMC Records.

53. T. J. Pietrini letter to Peter Fugina, May 11, 1961, Box 1, Folder 02.03.01.00000012, Cina Papers.

54. Transcript of testimony to Minnesota House Committee on Taxes, regarding HF 116, Taconite Constitutional Amendment, St. Paul, May 8–9, 1961, 8, in Taxes Equalization Program 1961 Folder, Box 1987.0970.MSS.00000024, BBHMC Records; Robert Linney letter to Reserve Employees, September 12, 1962, Box 1, Folder 02.03.01.00000013, Cina Papers.

55. Newlund, "Senate Tax Unit Supports Taconite Equity Guarantee"; Newlund, "Iron Officials."

56. "Taconite Proposal Hearings Start Today," *Minneapolis Tribune*, May 8, 1961.

57. "House Unit Ties Up Taconite Tax Bill," *Minneapolis Tribune*, April 14, 1961; Newlund, "Vote Kills Taconite Tax Plan"; "Minnesota Refuses New Tax Aid for Taconite Plants," *New York Times*, May 31, 1961.

58. John C. McDonald, "State Jaycees Start Petition Campaign for Taconite Amendment," *Minneapolis Tribune*, August 12, 1961; Frank Bourgin letter to Fred Cina, December 17, 1961, Box 1, Folder 02.03.01.00000012, Cina Papers; Mercer Cross, "Andersen Calls for Taconite Aid Action," *Minneapolis Tribune*. August 4, 1961.

59. "The DFL and Taconite" editorial, *Minneapolis Tribune*, October 24, 1961; "Andersen Asks Referendum on Taconite Tax," *Minneapolis Tribune*, May 18, 1961.

60. "'62 Issues Drawn for Minnesotans," *New York Times*, October 29, 1961.

61. "Andersen Calls Cina Taconite Tax Plan 'Encouraging,'" *Minneapolis Tribune*, February 10, 1962; "DFL Urged to Back Ore Amendment," *Minneapolis Tribune*, February 14, 1962; "DFL Stand on Taconite Called 'Stab in the Back,'" *Minneapolis Tribune*, March 10, 1962.

62. Earl T. Bester form letter, August 17, 1962, Correspondence, 1911–1973, Folder 3, Box 1, EWD Papers; "Rebuff to Taconite" editorial, *Minneapolis Tribune*, September 6, 1962; Dublin and Licht, *The Face of Decline*, 113.

63. Charles L. Horn, "Who Is Crying: 'Wolf'?" press release, February 19, 1962, Correspondence, 1911–1973, Folder 3, Box 1, EWD Papers; "70% Back Taconite Proposal," *Minneapolis Tribune*, March 4, 1962; "Governor Cites Poll on Taconite," *Minneapolis Tribune*, March 5, 1962; "Theory and Taconite" editorial, *Minneapolis Tribune*, April 12, 1962.

64. Sam Romer, "How the Accord Was Reached," *Minneapolis Tribune*, February 2, 1963; Mercer Cross, "Both Factions Agree on Need," *Minneapolis Tribune*, February 2, 1963; Dick Cunningham, "Miners Grudgingly Give Support to Taconite Amendment," *Minneapolis Tribune*, February 24, 1963; Mercer Cross, "Cautious Attitude on Taconite Held by DFL Leaders," *Minneapolis Tribune*, February 3, 1963. Labor reporters at the time credited the labor–management accord on the taconite amendment to the new cooperative bargaining process, called the Human Relations Committee, adopted by the steel industry (A. H. Raskin, "Nonstop Talks Instead of Nonstop Strikes," *New York Times*, July 7, 1963).

65. Cross, "Cautious Attitude on Taconite Held by DFL Leaders"; Mercer Cross, "DFL Is Preparing a Compromise on Taconite Proposal," *Minneapolis Tribune*, March 3, 1963; Donald D. Wozniak, transcript of speech to USWA local 1663, Hibbing, Minnesota, November 26, 1963, in Taconite Amendment 1961 Folder, Box 1987.0970.MSS.00000127, BBHMC Records; Mercer Cross, "Forsythe Asks DFL 'Get in Step,' Back Taconite Tax Amendment," *Minneapolis Tribune*, February 4, 1963.

66. "United States Steel Steps Up Planning for Taconite Plant," *New York Times*, August 30, 1963; "Constitutional Balm for an Old Sore," 112; Mercer Cross, "House Unit Approves 2 Taconite Tax Bills," *Minneapolis Tribune*, March 12, 1963.

67. Mercer Cross, "Action on Taconite Expected in House," *Minneapolis Tribune*, March 10, 1963; Al McConagha, "Andersen Signs Bill Providing Tax Equity on Taconite," *Minneapolis Tribune*, March 19, 1963; Mercer Cross, "House Unit Approves 2 Taconite Tax Bills," *Minneapolis Tribune*, March 12, 1963; Mercer Cross, "Taconite Bills Easily Clear House," *Minneapolis Tribune*, March 15, 1963; Al McConagha, "State Senate Passes Equity Amendment on Taconite Taxes," *Minneapolis Tribune*, March 21, 1963; "Taconite Measure Is Passed," *Minneapolis Tribune*, March 22, 1963.

68. Sam Romer, "Steelworker Chief Supports Taconite Proposal 'Fully,'" *Minneapolis Tribune*, October 22, 1963; Steelworkers Committee for Taconite Amendment, "Questions and Answers concerning Taconite Amendment (No. 1)," undated pamphlet, in Taconite Amendment publicity Folder, Box 2, EWD Papers; "Taconite Backers Map Final Effort," *Duluth News-Tribune*, October 11, 1964.

69. Bigart, "Poverty Blights Iron Ore Region"; Emil Pesonen, letter to the editor, *Mesabi Daily News*, October 22, 1964.

70. "Taconite Group Has Bipartisan Executive Unit," *Minneapolis Tribune*, January 19, 1964.

71. Richard P. Kleeman, "Many State Public Schools Issue Materials on Taconite Amendment," *Minneapolis Tribune*, October 18, 1964; "Taconite Backers Map Final Effort."

72. Rita Shemesh memorandum to Citizens' Committee for the Taconite Amendment, July 30, 1964, in Misc. folder, Citizens' Committee for the Taconite Amendment Papers, Minnesota Historical Society, St. Paul, Minnesota; Sam Romer, "Taconite Proposal Wins Approval," *Minneapolis Tribune*, November 4, 1964. See also DFL Party folder, Citizens' Committee for the Taconite Amendment papers, MHS; Frank Premack, "DFL Backs Taconite Proposal," *Minneapolis Tribune*, June 28, 1964; Leonard Inskip, "Amendment Would Unlock State's Taconite Resources," *Minneapolis Tribune*, October 27, 1964; Leonard Inskip, "'Immature' Mining Industry Needs Capital," *Minneapolis Tribune*, October 28, 1964; Leonard Inskip, "Taconite Tax Proposal Results from State's Loss of Iron Ore 'Monopoly,'" *Minneapolis Tribune*, October 29, 1964; Leonard Inskip, "Amendment Treats Taconite Industry as 'Manufacturing,'" *Minneapolis Tribune*, October 30, 1964; and Leonard Inskip, "State's Stake in Taconite Huge," *Minneapolis Tribune*, October 31, 1964.

73. Mercer Cross, "5 Liberals Form Group to Oppose Taconite Plan," *Minneapolis Tribune*, February 12, 1964; Bob Weber, "Taconite Foes Lack Publicity?" *Minneapolis Star*, October 19, 1964; "Taconite Plan Is Debated by Cina, Grittner," *Minneapolis Tribune*, February 28, 1964.

74. Romer, "Taconite Proposal Wins Approval"; Robert J. O'Keefe, "4-Year Drive for Taconite Vote Ending," *St. Paul Pioneer Press*, November 1, 1964; Ron Waataja, "Taconite Proposal Ahead," *St. Paul Pioneer Press*, November 4, 1964; "Taconite Wins a Tremendous Victory," editorial, *Minneapolis Tribune*, November 5, 1964; "Taconite Success" editorial, *St. Paul Pioneer Press*, November 5, 1964.

75. "Mining: The Iron Amendment," 74; "Production: A Pellet Gives Iron Ore Industry Shot in the Arm," *Business Week*, December 4, 1965, 107, 110. On the increase in iron-ore pelletizing capacity during these years, see Gerald Manners, *The Changing World Market for Iron Ore, 1950–1980: An Economic Geography* (Baltimore: Johns Hopkins University Press, 1971), 162–63.

76. Austin C. Wehrwein, "Economy Reversed in Mesabi Range," *New York Times*, November 21, 1965; Austin C. Wehrwein, "New Boom in Ore Stirs Minnesota," *New York Times*, May 23, 1965; "Mining: The Iron Amendment," 74.

77. "Mining: The Iron Amendment," 74; Gayle Anderson, "Natural Ores to Taconite—A Transitional Emergency," student essay reprinted in *Iron Ore and Men*, May 1963, 16, in Publications 1934–1963 Folder 3, Box 4, EWD Papers. *Iron Ore and Men* was the company newsletter for the Oliver Mining Company.

78. IRRRB, *Biennial Report, 1966–1968* (St. Paul: IRRRB, 1968); "Production: A Pellet Gives Iron Ore Industry Shot in the Arm," 112, 114; Horace T.

Reno and Helen E. Lewis, "Iron Ore," in *Minerals Yearbook 1957* (Washington, D.C.: U.S. Bureau of Mines, 1958), 584.

79. See Brown, *Overburden*, 45; Clements, *After the Boom in Tombstone and Jerome*, 2–3; and Mary Murphy, *Mining Cultures: Men, Women, and Leisure in Butte, 1914–1941* (Urbana: University of Illinois Press, 1997), xiii, as well as the critique of boom-and-bust rhetoric in the introduction.

80. "Crosby, Minn.," *New York Times*, March 26, 1967; Fillmore C. F. Earney, "New Ores for Old Furnaces: Pelletized Iron," *Annals of the Association of American Geographers* 59, no. 3 (1969): 531. On the Michigan iron ranges, see Reynolds and Dawson, *Iron Will*.

81. Zelizer, "The Uneasy Relationship," 282.

82. Brown, *Overburden*, 185.

83. On the labor–liberal alliance, see Battista, *The Revival of Labor Liberalism*, 1–26; and Boyle, *The UAW and the Heyday of American Liberalism, 1945–1968*.

3. TACONITE BITES BACK

1. "Pollution: The Classic Case," *Time*, May 6, 1974, http://www.time.com/time/magazine/article/0,9171,943689,00.html.

2. Tenner, *Why Things Bite Back*. See also Timothy LeCain, "When Everybody Wins Does the Environment Lose? The Environmental Techno-Fix in Twentieth-Century American Mining," in *The Technological Fix: How People Use Technology to Create and Solve Problems*, ed. Lisa Rosner (New York: Routledge, 2004), 137–53.

3. Davis, *Pioneering with Taconite*, 125–26; Thomas R. Huffman, "Exploring the Legacy of Reserve Mining: What Does the Longest Environmental Trial in History Tell Us about the Meaning of American Environmentalism?" *Journal of Policy History* 12, no. 3 (2000): 340–41.

4. Davis, *Pioneering with Taconite*, 43; William Appleby, "Minnesota Mines Experiment Station," *Minnesota Techno-Log*, March 1926, 181; Frank Schaumburg, *Judgment Reserved: A Landmark Environmental Case* (Reston, Va.: Reston Publishing, 1976), 214–15.

5. See Joel Mokyr, *The Lever of Riches: Technological Creativity and Economic Progress* (New York: Oxford University Press, 1990), 34–35, 90–92.

6. Smith, *Mining America*, 9–10, 57.

7. Davis, *Pioneering with Taconite*, 127–28; E. W. Davis, "Pollution," unpublished manuscript (1971–72), 7–8, in Lake Studies Folder, Box 1, EWD Papers; Thomas F. Bastow, *"This Vast Pollution . . .": United States of America v. Reserve Mining Company* (Washington, D.C.: Green Fields, 1986), 33.

8. Schaumburg, *Judgment Reserved*, 46, 54; Davis, "Pollution"; Bastow, *"This Vast Pollution . . . ,"* 5, 7; Robert V. Bartlett, *The Reserve Mining Controversy: Science, Technology, and Environmental Quality* (Bloomington: Indiana University Press, 1980), 22–23; Davis, *Pioneering with Taconite*, 135.

9. Schaumburg, *Judgment Reserved*, 55–56; Bartlett, *The Reserve Mining Controversy*, 25.

10. Bartlett, *The Reserve Mining Controversy*, 27–28.

11. Samuel P. Hays, *Beauty, Health, and Permanence: Environmental Poli-*

tics in the United States, 1955–1985 (New York: Cambridge University Press, 1987); Carolyn Merchant, *The Columbia Guide to American Environmental History* (New York: Columbia University Press, 2002); J. R. McNeill, *Something New under the Sun: An Environmental History of the Twentieth-Century World* (New York: W. W. Norton, 2000); Paul C. Milazzo, *Unlikely Environmentalists: Congress and Clean Water, 1945–1972* (Lawrence: University Press of Kansas, 2006), 3–10.

12. Hays, *Beauty, Health, and Permanence*, 78, 153; Bastow, *"This Vast Pollution . . . ,"* 26.

13. Milazzo, *Unlikely Environmentalists*, 3; Schaumburg, *Judgment Reserved*, 60–61; Merchant, *The Columbia Guide to American Environmental History*, 180–81, 245; Hays, *Beauty, Health, and Permanence*, 78–79, 198–99.

14. Milazzo, *Unlikely Environmentalists*, ix–x.

15. John Blatnik, "Need for Environmental Balance and the Environmental Balance Association of Minnesota, Inc.," transcript of speech, December 10, 1975, Personal Subject Files: Environmental Balance Assn of Minn. Inc. Folder, Box 108, Blatnik Papers, Minnesota Historical Society, St. Paul (hereafter Blatnik Papers).

16. Milazzo, *Unlikely Environmentalists*, 28.

17. Ibid., 23–24.

18. John Blatnik, "Political Speech for Station WMFG," August 15, 1940, 2, Political Files: Campaign 1940 Folder, Box 114, Blatnik Papers.

19. David Zwick and Marcy Benstock, *Water Wasteland: Ralph Nader's Study Group Report on Water Pollution* (New York: Grossman, 1971), 144–45; Bastow, *"This Vast Pollution . . . ,"* 8.

20. Schaumburg, *Judgment Reserved*, 60; Izaak Walton League, "Public Information and Conservation Education Honor Roll," June 25, 1960, Izaak Walton League of America Inc. Folder, Box 4, Blatnik Papers; Hubert Humphrey to John Blatnik, July 28, 1964, Personal Correspondence January 24–April 20, 1964, Folder, Box 107, Blatnik Papers.

21. Milazzo, *Unlikely Environmentalists*, 21–23.

22. Ibid., 30–35.

23. See Duluth Water Quality Control Laboratory Folder, Box 4, Blatnik Papers; Bastow, *"This Vast Pollution . . . ,"* 1–2; Zwick and Benstock, *Water Wasteland*, 155; Bartlett, *The Reserve Mining Controversy*, 53.

24. Bartlett, *The Reserve Mining Controversy*, 41, 56; Huffman, "Exploring the Legacy of Reserve Mining," 354; Grant Merritt, transcript of oral history interview, Margaret Robinson, interviewer, 1988, 20, in Minnesota Environmental Issues Oral History Project, Minnesota Historical Society, St. Paul; Bastow, *"This Vast Pollution . . . ,"* 28–29; Zwick and Benstock, *Water Wasteland*, 153.

25. Schaumburg, *Judgment Reserved*, 46, 54; Bastow, *"This Vast Pollution . . . ,"* 8; Bartlett, *The Reserve Mining Controversy*, 54.

26. Bartlett, *The Reserve Mining Controversy*, 54, 57–62; Bastow, *"This Vast Pollution . . . ,"* 14–18; Schaumburg, *Judgment Reserved*, 66, 68; Huffman, "Exploring the Legacy of Reserve Mining," 341; Gladwin Hill, "U.S. Report Cites Ore Plant Pollution," *New York Times*, May 11, 1969.

27. Bastow, *"This Vast Pollution . . . ,"* 18–21; Bartlett, *The Reserve Mining Controversy,* 59–63; Schaumburg, *Judgment Reserved,* 67. Secretary Udall later denied that Blatnik "intervened or attempted to influence . . . or to 'suppress' any part of the Department's report on Lake Superior" ("Blatnik Water Pollution Record Set Straight," *Congressional Record* 115 [July 2, 1969]: 18338).

28. Bastow, *"This Vast Pollution . . . ,"* 20–21; Schaumburg, *Judgment Reserved,* 64.

29. Milazzo, *Unlikely Environmentalists,* 54; Schaumburg, *Judgment Reserved,* 62–63; Bastow, *"This Vast Pollution . . . ,"* 24–25.

30. Schaumburg, *Judgment Reserved,* 74; Wendy Adamson, *Saving Lake Superior: A Story of Environmental Activism* (Minneapolis: Dillon Press, 1974), 50; Bastow, *"This Vast Pollution . . . ,"* 32, 42; Bartlett, *The Reserve Mining Controversy,* 75, 77–80.

31. Bastow, *"This Vast Pollution . . . ,"* 36–39, 44; Gladwin Hill, "Ore Company Defends Dumping of Its Waste into Lake Superior," *New York Times,* May 15, 1969; Schaumburg, *Judgment Reserved,* 87.

32. Bastow, *"This Vast Pollution . . . ,"* 45–47; Schaumburg, *Judgment Reserved,* 115–27; Bartlett, *The Reserve Mining Controversy,* 94–97.

33. Bastow, *"This Vast Pollution . . . ,"* 47–55; Schaumburg, *Judgment Reserved,* 137–38; Seth S. King, "Offer to Curb Lake Superior Dumping Fails to Satisfy Two States," *New York Times,* January 17, 1971; Reserve Mining Company, *Plan to Modify Tailings Discharge System* (Babbitt and Silver Bay, Minn.: Reserve Mining Company, 1971), in Box 2, EWD Papers.

34. Bastow, *"This Vast Pollution . . . ,"* 38–41, 58–61; Schaumburg, *Judgment Reserved,* 143–45.

35. Daniel A. Farber, "Risk Regulation in Perspective: Reserve Mining Revisited," *Environmental Law* 21, no. 4 (1991): 1322; Huffman, "Exploring the Legacy of Reserve Mining," 343–44; Saul Friedman, "But Nobody's Watching," *Chicago Tribune,* August 20, 1973.

36. Bastow, *"This Vast Pollution . . . ,"* 48–50.

37. Ibid., 68–73; Schaumburg, *Judgment Reserved,* 147–48.

38. Bastow, *"This Vast Pollution . . . ,"* 73–75, 141; "Superior, Private Dump?" *New York Times,* July 7, 1974; Farber, "Risk Regulation in Perspective," 1327.

39. Bastow, *"This Vast Pollution . . . ,"* 89, 113.

40. Ibid., 94–95; Bartlett, *The Reserve Mining Controversy,* 119; Huffman, "Exploring the Legacy of Reserve Mining," 341; Schaumburg, *Judgment Reserved,* 173n3.

41. Cummingtonite-grunerite is defined as "a general name for a series or group of amphibole minerals that are essentially identical except for the relative quantities of iron and magnesium in them. The more iron-rich members are sometimes referred to as grunerites, although the word cummingtonite is all inclusive and refers to the entire series" (Schaumburg, *Judgment Reserved,* 173n2); Bastow, *"This Vast Pollution . . . ,"* 96–101, 112; Bartlett, *Reserve Mining Controversy,* 122–23; Schaumburg, *Judgment Reserved,* 149–50; Jane E. Brody, "Asbestos Found in Duluth Water," *New York Times,* June 16, 1973.

42. Bastow, *"This Vast Pollution . . . ,"* 129, 121.

43. Ibid., 91–93; Schaumburg, *Judgment Reserved*, 228; Paul Brodeur, *Expendable Americans* (New York: Viking, 1974).

44. Bartlett, *The Reserve Mining Controversy*, 126–29; Jane E. Brody, "Iron Ore Company vs. the Changing Times: U.S. Court Will Decide on Right to Pollute," *New York Times*, August 8, 1973; Thomas R. Huffman, "Enemies of the People: Asbestos and the Reserve Mining Trial," *Minnesota History* 59, no. 7 (2005): 299; Wade Green, "Life vs. Livelihood," *New York Times Magazine*, November 24, 1974.

45. Bartlett, *The Reserve Mining Controversy*, 128; Luther J. Carter, "Pollution and Public Health: Taconite Case Poses Major Test," *Science* 186, no. 4158 (1974): 34.

46. Casey Bukro, "2 Million Asbestos Fibers per Gulp—Duluth Up in Arms," *Chicago Tribune*, April 2, 1975; William E. Farrell, "Duluth Stoically Awaits Tests for Cancer Risk in Its Water," *New York Times*, June 24, 1973.

47. Bastow, *"This Vast Pollution . . . ,"* 104–6.

48. Ibid., 114–19.

49. Ibid., 142–47; William E. Farrell, "Trial on the Discharge of Ore Wastes into Lake Superior Is Nearing Completion," *New York Times*, March 19, 1974.

50. Bastow, *"This Vast Pollution . . . ,"* 149–51.

51. Bartlett, *The Reserve Mining Controversy*, 164–65, 172; Schaumburg, *Judgment Reserved*, 221; Bastow, *This Vast Pollution*, 151, 155–57.

52. Bastow, *"This Vast Pollution . . . ,"* 153–55.

53. Ibid., 158–61; Schaumburg, *Judgment Reserved*, 164–67.

54. Bastow, *"This Vast Pollution . . . ,"* 153, 163–71; Schaumburg, *Judgment Reserved*, 194; "Rule Dumping of Asbestos No Immediate Health Peril," *Chicago Tribune*, June 5, 1974.

55. Bastow, *"This Vast Pollution . . . ,"* 172–76.

56. Ibid., 180–85.

57. Schaumburg, *Judgment Reserved*, 222–23.

58. Bartlett, *The Reserve Mining Controversy*, 188–89, 198–99, 213–14; "Judge Bans Use of Lake for Minnesota Ore Waste," *New York Times*, July 8, 1976; "Reserve Mining Wins Court Battle over Disposal Site for Ore Wastes," *New York Times*, April 9, 1977.

59. Bartlett, *The Reserve Mining Controversy*, 168, 206; Bastow, *"This Vast Pollution . . . ,"* 192; "Reserve Mining Ends Dumping in Lake," *New York Times*, March 18, 1980.

60. "Closing of Mining Plant Brings Shock and Dismay," *Chicago Tribune*, April 22, 1974.

61. E. C. Lamfman, *Reserve Mining Company Newsletter,* January 23, 1969, in Box 2, EWD Papers; William Verity, "Armco Defends Itself," *Chicago Tribune*, April 21, 1975.

62. Hays, *Beauty, Health, and Permanence*, 307–15; Bartlett, *The Reserve Mining Controversy*, 8–9.

63. "Closing of Mining Plant Brings Shock and Dismay."

64. "Workers Rejoice as Mine Reopens," *Chicago Tribune*, April 24, 1974.

65. Bartlett, *The Reserve Mining Controversy,* 128, 160; "Minnesota Town Prevails—For Now," *Chicago Tribune,* October 13, 1975.

66. Harvey D. Shapiro, "Taconite Rejuvenates the Iron Range," *New York Times,* August 3, 1975.

67. Bastow, *"This Vast Pollution . . . ,"* 109; Bartlett, *The Reserve Mining Controversy,* 194.

68. Arnold R. Alanen, "Morgan Park: U.S. Steel and a Planned Company Town," in *Duluth: Sketches of the Past,* ed. Ryck Lydecker and Lawrence Sommer (Duluth: Northprint, 1976), 123–34; Alanen, "Years of Change on the Iron Range," 187. See also Arnold Alanen, *Morgan Park: Duluth, U.S. Steel, and the Forging of a Company Town* (Minneapolis: University of Minnesota Press, 2007).

69. Brian K. Obach, *Labor and the Environmental Movement: The Quest for Common Ground* (Cambridge: MIT Press, 2004), 84; Robert Gordon, "'Shell No!' OCAW and the Labor-Environmental Alliance," *Environmental History* 3, no. 4 (1998): 462.

70. Hays, *Beauty, Health, and Permanence,* 429; Bartlett, *The Reserve Mining Controversy,* 223.

71. "Minnesota Town Prevails—For Now"; "Silver Bay: Living in Limbo," *Time,* January 10, 1977, http://www.time.com/time/magazine/article/0,9171,712342,00.html.

72. David Rosner and Gerald Markowitz, *Deadly Dust: Silicosis and the Ongoing Struggle to Protect Workers' Health,* 2d ed. (Ann Arbor: University of Michigan Press, 2006), xvii.

73. Scott Dewey, "Working for the Environment: Organized Labor and the Origins of Environmentalism in the United States, 1948–1970," *Environmental History* 3, no. 1 (1998): 45–46, 50, 54; Gordon, "'Shell No!'" 464–67.

74. Chad Montrie, "Expedient Environmentalism: Opposition to Coal Surface Mining and the United Mine Workers of America, 1945–1975," *Environmental History* 5, no. 1 (2000): 78; Harry M. Caudill, *Night Comes to the Cumberlands: A Biography of a Depressed Area* (Boston: Little, Brown, 1963).

75. Dewey, "Working for the Environment," 45–46, 58; Obach, *Labor and the Environmental Movement,* 53.

76. Huffman, "Exploring the Legacy of Reserve Mining," 346.

77. Bastow, *"This Vast Pollution . . . ,"* 6–8.

78. Andrew Biemiller to John Blatnik, July 21, 1964, AFL–CIO Folder, Box 1, Blatnik Papers; Lisa M. Fine, "Rights of Men, Rights of Passage: Hunting and Masculinity at REO Motors of Lansing, Michigan, 1945–1975," *Journal of Social History* 33, no. 4 (2000): 807.

79. R. J. Thomas to John H. Lyons, June 30, 1973, Personal Cor. & Mis. by Individual: R. J. Thomas re: Reserve Mining Folder, Box 108, Blatnik Papers. This folder contains several such letters and information; Carter, "Pollution and Public Health," 36.

80. Gene A. Roach to John A. Blatnik, August 3, 1973, Lake Superior Asbestos Folder 3, Box 97, Blatnik Papers; R. J. Thomas to Hubert H. Humphrey, September 23, 1973, Personal Cor. & Mis. by Individual: R. J. Thomas re: Reserve Mining Folder, Box 108, Blatnik Papers.

81. Carl Hennemann, "He Hopes Amendment Will Pass—Dr. Edward Davis: 'Mr. Taconite,'" *St. Paul Dispatch,* October 21, 1964.

82. See, for example, the unpublished manuscript Davis wrote about his experiences in northern Minnesota: E. W. Davis, "The Walleyes and Val," unpublished manuscript, September 28, 1949, in Box 4, EWD Papers.

83. Davis, *Pioneering with Taconite,* 197; Hennemann, "He Hopes Amendment Will Pass"; "Ore-able Fate? He'd Treat Old Cars like Dirt," *St. Paul Pioneer Press,* January 20, 1968.

84. Smith, *Mining America,* 42–46; Manuel, "Mr. Taconite," 337–40.

85. E. W. Davis, letter to Minnesota Pollution Control Agency, November 16, 1972, Lake Studies Folder, Box 1, EWD Papers; Davis, "Pollution," 8–9; Huffman, "Enemies of the People," 296.

86. Jane Davis Wall to John Blatnik, June 26, 1973, Lake Superior Asbestos Folder 3, Box 97, Blatnik Papers.

87. John Blatnik to Mrs. Malcolm D. Wall, August 7, 1973, Lake Superior Asbestos Folder 3, Box 97, Blatnik Papers.

88. Bartlett, *The Reserve Mining Controversy,* 227.

89. Schaumburg, *Judgment Reserved,* 172.

90. Bartlett, *The Reserve Mining Controversy,* 10, 208; Jessica Steeno, "Cleanup Delays Residence Hall Project," *Minnesota Daily,* February 14, 1997; David Shaffer, "State Kept Quiet on Cancer in 35 Miners," *Minneapolis Star Tribune,* June 16, 2007; John Myers, "Links between Mesothelioma, Taconite Mining Hard to Find," *Duluth News Tribune,* May 31, 2012.

91. Grant Merritt transcript of oral history interview, 22.

92. Bartlett, *The Reserve Mining Controversy,* xii; "Asbestos on the North Shore?" *Chicago Tribune,* September 10, 1973.

93. Huffman, "Exploring the Legacy of Reserve Mining," 342, 346; "Reserve Closure," *Mining Journal,* June 18, 1982, 457.

4. ECONOMIC DEVELOPMENT POLICY, REGIONAL AND LOCAL

1. David M. Kennedy, *Freedom from Fear: The American People in Depression and War, 1929–1945* (New York: Oxford University Press, 1999), 246; Lyndon B. Johnson, "Remarks at the University of Michigan," May 22, 1964, online by Gerhard Peters and John B. Woolley, *The American Presidency Project,* www.presidency.ucsb.edu/ws/?pid=26262. There is a vast literature on the history of the federal welfare state in the twentieth-century United States. For useful overviews, see Edward Berkowitz, *America's Welfare State: From Roosevelt to Reagan* (Baltimore: Johns Hopkins University Press, 1991); Kessler-Harris, *In Pursuit of Equity*; and James T. Patterson, *America's Struggle against Poverty, 1900–1980* (Cambridge: Harvard University Press, 1982).

2. Alan Brinkley, *The End of Reform: New Deal Liberalism in Recession and War* (New York: Vintage, 1996); Otis L. Graham Jr., *Losing Time: The Industrial Policy Debate* (Cambridge: Harvard University Press, 1992). On postwar Keynesianism, see Robert Collins, *The Business Response to Keynes, 1929–1964* (New York: Columbia University Press, 1981); and Robert Collins, *More: The Politics of Economic Growth in Postwar America* (New York: Oxford University Press, 2002).

3. Gregory S. Wilson, *Communities Left Behind: The Area Redevelopment Administration, 1945–1965* (Knoxville: University of Tennessee Press, 2009). On economic development in postwar urban America, see Colin Gordon, *Mapping Decline: St. Louis and the Fate of the American City* (Philadelphia: University of Pennsylvania Press, 2009); Guian A. McKee, *The Problem of Jobs: Liberalism, Race, and Deindustrialization in Philadelphia* (Chicago: University of Chicago Press, 2008); Robert Self, *American Babylon: Race, Power, and the Struggle for the Postwar City in California* (Princeton, N.J.: Princeton University Press, 2003); and Sugrue, *The Origins of the Urban Crisis.* For examples of economic development policy in rural areas, see Ronald D. Eller, *Uneven Ground: Appalachia since 1945* (Lexington: University of Kentucky Press, 2008); and Bruce Schulman, *From Cotton Belt to Sunbelt: Federal Policy, Economic Development, and the Transformation of the South, 1938–1980* (Durham, N.C.: Duke University Press, 1994). On international development schemes, see Nils Gilman, *Mandarins of the Future: Modernization Theory in Cold War America* (Baltimore: Johns Hopkins University Press, 2007). There is also an extensive social-scientific literature on state-level economic development programs. See, for example, Peter Eisinger, "Do the American States Do Industrial Policy?" *British Journal of Political Science* 20, no. 4 (1990): 509–35; Peter K. Eisinger, *The Rise of the Entrepreneurial State: State and Local Economic Development Policy in the United States* (Madison: University of Wisconsin Press, 1988); and J. Craig Jenkins, Kevin T. Leicht, and Heather Wendt, "Class Forces, Political Institutions, and State Intervention: Subnational Economic Development Policy in the United States, 1971–1990," *American Journal of Sociology* 111, no. 4 (2006): 1122–80.

4. The IRRRB had several official titles over the years, including the Department of Iron Range Resources and Rehabilitation and Iron Range Resources. For simplicity, I use the term IRRRB when referring to the agency throughout this chapter. On the Iron Range, the agency is typically called the "I-Triple R-B."

5. Julian Zelizer, *Taxing America: Wilbur D. Mills, Congress, and the State, 1945–1975* (New York: Cambridge University Press, 1998). For an excellent overview of the field of policy history, see Meg Jacobs and Julian E. Zelizer, "The Democratic Experiment: New Directions in American Political History," in *The Democratic Experiment: New Directions in American Political History,* ed. Meg Jacobs, William J. Novak, and Julian E. Zelizer (Princeton, N.J.: Princeton University Press, 2003), 1–19.

6. Hickok, *One Third of a Nation,* 129–30.

7. Dana Miller, "Public Policy and Economic Development: The Iron Range Experience," paper presented at the Minnesota Historical Society Annual Meeting (Minneapolis, 1990), 4.

8. Fred A. Cina to President Franklin Delano Roosevelt, May 14, 1941, Box 1, Folder 01.02.00000003, Cina Papers.

9. President Franklin Roosevelt to Fred Cina, May 24, 1941, Box 1, Folder 01.02.00000003, Cina Papers. On the broader shift from domestic recovery to international issues at the time, see Brinkley, *The End of Reform,* 175–200, and Kennedy, *Freedom from Fear,* 381–515.

10. Edward G. Bayuk, *Iron Range Resources and Rehabilitation: Report to*

the *Governor and the Legislature for the Twenty-second Biennium, July 1, 1948–July 1, 1950* (St. Paul: IRRRB, 1950), 7; Dana H. Miller, *The Iron Range Resources and Rehabilitation Board: The First Fifty Years* (Eveleth, Minn.: IRRRB, 1991), 1; IRRRB, *The Iron Range Resources and Rehabilitation Board: 50 Years of Vision* (Eveleth, Minn.: IRRRB, 1991), 3.

11. IRRRB, *The Iron Range Resources and Rehabilitation Board*, 1; Margaret E. Dewar, "Development Analysis Confronts Politics: Industrial Policy on Minnesota's Iron Range," *Journal of the American Planning Association* 52, no. 3 (1986): 291; Miller, *The Iron Range Resources and Rehabilitation Board*, 1.

12. IRRRB, *Biennial Report, 1960–1962* (St. Paul: Department of Iron Range Resources and Rehabilitation, 1962), 3.

13. IRRRB, *Biennial Report, 1964–1966* (St. Paul: Department of Iron Range Resources and Rehabilitation, 1964), 6.

14. "Industrial Survey of Northeastern Minnesota," undated, Box 1, Folder 01.04.00000005, Cina Papers. As political scientist Peter Eisinger notes, similar critiques of industrial policy were raised nationwide during the postwar decades (Eisinger, "Do the American States Do Industrial Policy?" 510).

15. Wilson, *Communities Left Behind*, 1–29.

16. Edward G. Bayuk, *Developing Human and Natural Resources in Minnesota: A Report on the Work of the Iron Range Resources and Rehabilitation Commission, 1950–1952* (St. Paul: IRRRB, 1952), 5–6; Charles L. Horn, "The Range Resources Commission—A Potential Asset," memorandum, January 25, 1951, Box 10, Minneapolis General Folder, MES Records.

17. IRRRB, *Forward in Developing Natural Resources* (St. Paul: Iron Range Resources and Rehabilitation Commission, 1958), 8; IRRRB, *Search for Resources and New Uses of Them* (St. Paul: Iron Range Resources and Rehabilitation Commission, 1960), 7–8.

18. Bayuk, *Iron Range Resources and Rehabilitation*, 10–18.

19. On the history of logging in the upper Great Lakes region, see William Cronon, *Nature's Metropolis: Chicago and the Great West* (New York: W. W. Norton, 1991), 148–206; and Agnes M. Larson, *The White Pine Industry in Minnesota: A History* (1949; reprint, Minneapolis: University of Minnesota Press, 2007).

20. C. C. Crosby to Howard Siegel, December 30, 1940, Box 1, Folder 01.02.00000003, Cina Papers; Austin C. Wehrwein, "Labor, Business and Civic Leaders in the Mesabi Iron Range Join to Diversify Industry," *New York Times*, May 25, 1964; IRRRB, *Biennial Report, 1968–1970* (St. Paul: IRRRB, 1970), 5.

21. Bayuk, *Developing Human and Natural Resources in Minnesota*, 8–10; Perlstein, *Nixonland*, 329, 347. On the broader context of economic development policy, see Wilson, *Communities Left Behind*, 3.

22. Sar A. Levitan, *Federal Aid to Depressed Areas: An Evaluation of the Area Redevelopment Administration* (Baltimore: Johns Hopkins University Press, 1964), 1.

23. Quoted in Wilson, *Communities Left Behind*, 36.

24. Stein, *Running Steel, Running America*, 27; Wilson, *Communities Left Behind*, 32. See also Gregory S. Wilson, "Deindustrialization, Poverty, and Federal Area Redevelopment in the United States, 1945–1965," in Cowie and Heathcott, *Beyond the Ruins: The Meanings of Deindustrialization*, ed.

Jefferson Cowie and Joseph Heathcott (Ithaca, N.Y.: Cornell University Press, 2003), 189.

25. Wilson, *Communities Left Behind*, 31–56; Levitan, *Federal Aid to Depressed Areas*, viii, 6, 17, 20. On the 1960 presidential campaign, see Theodore H. White, *The Making of the President 1960* (New York: Atheneum, 1961).

26. Wilson, *Communities Left Behind*, 3; Wilson, "Deindustrialization, Poverty, and Area Redevelopment," 184–85. On Pennsylvania's state-level redevelopment agency, see Dublin and Licht, *The Face of Decline*, 114–35.

27. John A. Blatnik, interview by Joseph E. O'Connor, February 4, 1966, 1, 27–28, 31, John F. Kennedy Library Oral History Program, http://www.jfklibrary.org/Asset-Viewer/Archives/JFKOH-JOAB-01.aspx.

28. W. H. Lawrence, "Kennedy Sets Up Panel to Map Aid," *New York Times*, December 5, 1960; James Reston, "Washington," *New York Times*, March 25, 1962; IRRRB, *Biennial Report, 1960–1962*, 13.

29. IRRRB, *Biennial Report, 1964–1966* (St. Paul: Department of Iron Range Resources and Rehabilitation, 1966), 5; Wilson, "Deindustrialization, Poverty, and Area Redevelopment," 192–93.

30. Wilson, "Deindustrialization, Poverty, and Area Redevelopment," 192–93. On the revival of the industrial policy debate during the 1980s, see Eisinger, "Do the American States Do Industrial Policy?" 509–10; and Graham, *Losing Time*.

31. IRRRB, *Biennial Report, 1962–1964*, 2; "Minneapolis," *New York Times*, October 24, 1965; IRRRB, *Biennial Report, 1968–1970* (St. Paul: Department of Iron Range Resources and Rehabilitation, 1970), 4.

32. Harvey D. Shapiro, "Taconite Rejuvenates the Iron Range," *New York Times*, August 3, 1975; Center for Small Towns, University of Minnesota Morris, Historical Minnesota Census Data, www.morris.umn.edu/services/cst/dar/mdc/census/index.php.

33. State of Minnesota, Office of the Legislative Auditor, "Audit Report: Iron Range Resources and Rehabilitation Board," St. Paul, April 22, 1980, 4; see also Kent Curtis, "Greening Anacanda: EPA, ARCO, and the Politics of Space in Postindustrial Montana," in Cowie and Heathcott, *Beyond the Ruins*, 91–111.

34. IRRRB, *Biennial Report, 2001–2002* (Eveleth, Minn.: Iron Range Resources and Rehabilitation Agency, 2002), 10; IRRRB, *Biennial Report, 1970–1972*, 8. On the program's ongoing popularity, see IRRRB, *Biennial Report, 1976–1978* (St. Paul: Iron Range Resources and Rehabilitation Board, 1978), 7; and Brown, *Overburden*, 86. Although many residents supported the waste removal program, historic preservationists worried that historically significant buildings were being demolished.

35. IRRRB, *Biennial Report, 1970–1972*, 9, 34; IRRRB, *Search for Resources and New Uses of Them*; IRRRB, *Biennial Report, 1976–1978*, 4.

36. IRRRB, "Ely and Its Resources: 'The Friendly City,'" September 1959, 3, Community Histories, Ely, Minnesota, Collection, Iron Range Research Center, Chisholm, Minnesota.

37. "Final Report, Title I Project, Ely, Minnesota," January 1968–December

31, 1970, 1, 2, 16, Community Histories, Ely, Minnesota, Collection, Iron Range Research Center, Chisholm, Minnesota.

38. Charles Brubaker, Sam Caudill, and Northeast Minnesota Chapter of the American Institute of Architects, "Report from the American Institute of Architects' Urban Design Team Study of Ely, Minnesota," July 18–20, 1970, Community Histories, Ely, Minnesota, Collection. Iron Range Research Center, Chisholm, Minnesota; Aguar Jyring Whiteman Moser, Inc., "Gilbert Reconnaissance Survey," September 1971, 5, Community Histories, Gilbert, Minnesota, Collection, Iron Range Research Center, Chisholm, Minnesota; "Final Report, Title I Project, Ely, Minnesota," 17.

39. Vladimar Shipka, quoted in IRRRB Meeting Minutes, January 24, 1979, Eveleth, 12; State of Minnesota, Office of the Legislative Auditor, "Audit Report: Iron Range Resources and Rehabilitation Board," St. Paul, April 22, 1980, 2, 7–8. The IRRRB's grant activities were again criticized by the Office of the Legislative Auditor in 1985 (State of Minnesota, Office of the Legislative Auditor, Financial Audit Division, "Iron Range Resources and Rehabilitation Board: Statewide Audit Management Letter, Fiscal Year 1984," St. Paul, March 1985, MN DOCS Microfiche 85–0686).

40. IRRRB Meeting Minutes, November 9, 1981, Eveleth, 26; IRRRB, *Biennial Report, 1980–1982*, 34–35; Dewar, "Planning Analysis Confronts Politics," 292.

41. Dewar, "Planning Analysis Confronts Politics," 292–94.

42. Tony Kennedy, "High-Tech Chopsticks Factory Hopes to Make a Fortune, Cookie," Associated Press, December 9, 1986, www.lexisnexis.com.

43. Frances Phillips, "Chopping His Way into Eastern Markets," *Financial Post* (Toronto), November 15, 1988, www.lexisnexis.com; Kennedy, "High-Tech Chopsticks Factory Hopes to Make a Fortune, Cookie"; Tony Kennedy, "Chopsticks Plant Closed for Repairs," Associated Press, January 20, 1989, www.lexisnexis.com; "Chopsticks Factory Closes," Associated Press, July 19, 1989; "Chopsticks Factory Closes Again," United Press International, July 18, 1989, www.lexisnexis.com.

44. Norman Prahl, quoted in IRRRB Meeting Minutes, January 24, 1979, Eveleth, 17.

45. Dewar, "Development Analysis Confronts Politics," 295.

46. IRRRB, *Biennial Report, 1982–1984* (St. Paul: IRRRB, 1984), 2.

47. Steven Greenhouse, "An Ore Carrier's Troubled Odyssey," *New York Times*, July 14, 1985; IRRRB, "Action Plan, 1984–86," Eveleth, June 1, 1984, 2, MN DOCS Microfiche 85–0632.

48. Arthur D. Little, Inc., "Analysis of Operations, Opportunities and Implementation Strategies for the Iron Range Resources and Rehabilitation Board," 1985, MN DOCS Microfiche 85–1190.

49. IRRRB Meeting Minutes, June 18, 1975, 40.

50. IRRRB, *87–88 Executive Summary* (Eveleth, Minn.: IRRRB, 1988), 4; IRRRB, *91/92 Biennial Report* (Eveleth, Minn.: IRRRB, 1992), 3. Miller, *The Iron Range Resources and Rehabilitation Board*, 15–16.

51. Robert A. Beauregard, "Constituting Economic Development: A Theoretical Perspective," in *Theories of Economic Development: Perspectives from*

across the Disciplines,ed. Richard D. Bingham and Robert Mier (Newbury Park, Calif.: SAGE, 1993), 278. On the celebration of the entrepreneur at this time, see Thomas Frank, *One Market under God: Extreme Capitalism, Market Populism, and the End of Economic Democracy* (New York: Anchor, 2000), 30–31, 80–81, 200.

52. D. J. Tice, "The Thing on the Hill (Part II)," *Minnesota Corporate Report*, November 1982, 98–99.

53. For an overview of some postwar development plans, see Graham, *Losing Time*; Dublin and Licht, *The Face of Decline*, 114–35; Joseph Heathcott and Máire Agnes Murphy, "Corridors of Flight, Zones of Renewal: Industry, Planning, and Policy in the Making of Metropolitan St. Louis, 1940–1980," *Journal of Urban History* 31, no. 2 (2005): 151–89; and Sugrue, *The Origins of the Urban Crisis*.

54. Thomas J. Sugrue, "All Politics Is Local: The Persistence of Localism in Twentieth-Century America," in Jacobs, Novak, and Zelizer, *The Democratic Experiment*, 304.

55. High, *Industrial Sunset*. On the political responses of industrial workers to deindustrialization, see Jefferson Cowie, *Stayin' Alive: The 1970s and the Last Days of the Working Class* (New York: New Press, 2010).

56. Beauregard, "Constituting Economic Development," 267.

5. THE TURN TO HERITAGE

1. Carlo Rotella, *Good with Their Hands: Boxers, Bluesmen, and Other Characters from the Rust Belt* (Berkeley: University of California Press, 2002), 9; Mike Wallace, *Mickey Mouse History and Other Essays on American Memory* (Philadelphia: Temple University Press, 1996), 92–93.

2. Rotella, *Good with Their Hands*, 197.

3. Cathy Stanton, "Performing the Postindustrial: The Limits of Radical History in Lowell, Massachusetts," *Radical History Review*, no. 98 (2007): 83–84. See also Cathy Stanton, *The Lowell Experiment: Public History in a Postindustrial City* (Amherst: University of Massachusetts Press, 2006); and Andrew Hurley, *Beyond Preservation: Using Public History to Revitalize Inner Cities* (Philadelphia: Temple University Press, 2010). On Minnesota's architectural preservation movement during these years, see Minnesota Historical Society and Minnesota State Planning Agency, *Historic Preservation for Minnesota Communities* (St. Paul: Minnesota State Planning Agency, 1980).

4. Stanton, "Performing the Postindustrial," 82.

5. Kathryn Marie Dudley, *The End of the Line: Lost Jobs, New Lives in Postindustrial America* (Chicago: University of Chicago Press, 1994), 49–100.

6. Rotella, *Good with Their Hands*, 168, 189.

7. Steven High and David W. Lewis, *Corporate Wasteland: The Landscape and Memory of Deindustrialization* (Ithaca, N.Y.: Cornell University Press, 2007), 41–42. Background on the industrial archaeology movement in the United States is documented in *IA: The Journal for the Society of Industrial Archeology*, which began publishing in 1975.

8. Peter Goin and C. Elizabeth Raymond, "Recycled Landscapes: Mining's Legacy on the Mesabi Iron Range," in *Technologies of Landscape: From*

Reaping to Recycling, ed. David E. Nye (Amherst: University of Massachusetts Press, 1999), 273, 271; Peter Goin and C. Elizabeth Raymond, *Changing Mines in America* (Santa Fe, N.Mex.: Center for American Places, 2004), xii, 7. On the interpretation of the mining and postmining landscape, see Richard V. Francaviglia, *Hard Places: Reading the Landscape of America's Historic Mining Districts* (Iowa City: University of Iowa Press, 1991).

9. Dudley, *The End of the Line,* xxiii, 175–78; High and Lewis, *Corporate Wasteland,* 25.

10. Landis, *Three Iron Mining Towns,* 51, 16–17. On upper Midwest tourism, see Aaron Shapiro, *The Lure of the North Woods: Cultivating Tourism in the Upper Midwest* (Minneapolis: University of Minnesota Press, 2013).

11. Clay Blair Jr., "Minnesota Grows Older," *Saturday Evening Post,* March 18, 1961, 85; Educational Research and Development Council of Northeast Minnesota, *Economy of Northeast Minnesota: Current Economic Activity* (Duluth, Minn.: Educational Research and Development Council of Northeast Minnesota, 1968), 33. A copy of this report is available at the Iron Range Research Center, Chisholm, Minnesota. Advertisement in Iron Range Enterprises, *The Mesabi-Vermilion Iron Range* (Biwabik, Minn.: Iron Range Enterprises, 1965), 1.

12. Shapiro, *The Lure of the North Woods,* xvi.

13. IRRRB, *Search for Resources and New Uses of Them,* 37; Midwest Research Institute (MRI), *Assessment of the Tourism Development Components of the Iron Range Interpretative Program* (Minnetonka, Minn.: MRI, 1980), 56. These reports are available at the Iron Range Research Center.

14. "Final Report, Title I Project, Ely, Minnesota," January 1968–December 31, 1970, 5, Community Histories: Ely, Minnesota, Collection, Iron Range Research Center, Chisholm, Minnesota; Educational Research and Development Council of Northeast Minnesota, *Economy of Northeast Minnesota,* 33.

15. Roger S. Williams, *Carey Lake Area Recreation Development: A Feasibility Study* (Duluth, Minn.: Aguar, Jyring and Whiteman Planning Associates, 1965), 1, 34–38, 4. The Carey Lake site eventually became a park maintained by the city of Hibbing.

16. Aguar Jyring Whiteman Moser, Inc., *Iron Range Interpretative Program: A Report to the Legislature* (Duluth, Minn.: Aguar Jyring Whiteman Moser, 1971), iii; *Minnesota's Iron Range Trail* (St. Paul: Minnesota Department of Conservation and Minnesota Department of Economic Development, undated brochure), 1; IRRRB, *Biennial Report, 1970–1972,* 34.

17. Midwest Research Institute, *Assessment of the Tourism Development Components of the Iron Range Interpretative Program,* 3–4.

18. Even after women were allowed into formerly male-only jobs in the mines, they faced harrowing discrimination and harassment on the job. See Clara Bingham and Laura Leedy Gansler, *Class Action: The Landmark Case That Changed Sexual Harassment Law* (New York: Anchor, 2002). On women entering the traditionally male field of industrial mining in the 1970s and 1980s, see Marat Moore, *Women in the Mines: Stories of Life and Work* (New York: Twayne, 1996).

19. Educational Research and Development Council of Northeast Minnesota, *Economy of Northeast Minnesota*, 38; D. J. Tice, "The Thing on the Hill (Part II)," *Minnesota Corporate Report*, November 1982, 96; Bigart, "Poverty Blights Iron Ore Region"; Bill Cook, oral history interview, Keewatin, Minnesota, August 9, 1986, Unemployment on the Iron Range Oral History Collection, A-86–694, Iron Range Research Center, Chisholm, Minnesota.

20. Brown, *Overburden*, 90; Thomas J. Baerwald, "Forces at Work on the Landscape," in *Minnesota in a Century of Change: The State and Its People since 1900*, ed. Clifford E. Clarke Jr. (St. Paul: Minnesota Historical Society Press, 1989), 19; Goin and Raymond, "Recycled Landscapes," 271; High and Lewis, *Corporate Wasteland*, 31. See also Kent C. Ryden, *Mapping the Invisible Landscape: Folklore, Writing, and the Sense of Place* (Iowa City: University of Iowa Press, 1993), 98–99.

21. IRRRB Meeting Minutes, February 27, 1973, St. Paul, 9; State Senator George Perpich, quoted in IRRRB Meeting Minutes, June 18, 1975, 45. On the creation of Voyageurs National Park, see Frederick Witzig, *Voyageurs National Park: The Battle to Create Minnesota's National Park* (Minneapolis: University of Minnesota Press, 2004).

22. Midwest Research Institute, *Assessment of the Tourism Development Components of the Iron Range Interpretative Program*, 7.

23. Aguar Jyring Whiteman Moser, Inc., "From Depot to Cultural Center," report, March 9, 1972, 21, Community Histories: Duluth, Minnesota, Collection, Iron Range Research Center, Chisholm, Minnesota. On the history of industrial museums, see Wallace, *Mickey Mouse History and Other Essays on American Memory*, 88–100.

24. Brown, *Overburden*, 134.

25. IRRRB, *Biennial Report, 1970–1972*, 34.

26. Ibid., 36.

27. IRRRB, *Biennial Report, 1972–1974*, 11.

28. "$650,000 Federal Funding Assures Construction of $2 Million Interpretative Center," *Chisholm Free Press*, April 14, 1974; IRRRB, *Biennial Report, 1974–1976*, 15; IRRRB Meeting Minutes from May 18, 1973, St. Paul, 28.

29. IRRRB, *Biennial Report, 1972–1974*, 11; IRRRB, *Biennial Report, 1974–1976*, 15; IRRRB, *Biennial Report, 1976–1978*, 9.

30. IRRRB, *Biennial Report, 1972–1974*, 11; IRRRB, *Biennial Report, 1976–1978*, 10.

31. IRRRB Meeting Minutes, March 26, 1975, 4; September 11, 1975, 19–27.

32. IRRRB Meeting Minutes, June 23, 1976, 5; May 20, 1976, 34–42; February 18, 1976; February 25, 1976, 27–30.

33. Norbert Arnold, quoted in IRRRB Meeting Minutes, March 3, 1976, 17–18; June 23, 1976, 2; July 1, 1976, 4.

34. IRRRB Meeting Minutes, November 26, 1979, 23–35; January 28, 1976, 55–57.

35. Midwest Research Institute, *Assessment of the Tourism Development Components of the Iron Range Interpretative Program*, 14–17, 48–49, 56.

36. IRRRB, *Biennial Report, 1976–1978*, 11–12.

37. Ibid., 12; IRRRB, *Biennial Report, 1978–1980*, 10.

38. Catherine Rukavina and Ron Dicklich, quoted in IRRRB Meeting Minutes, June 15, 1981, 43, 68.

39. "A 70 Foot High, Sixty Ton Monument to Iron Ore Miner Now Under Construction," *Chisholm Free Press*, April 15, 1976.

40. Kirk Savage, "Monuments of a Lost Cause: The Postindustrial Campaign to Commemorate Steel," in Cowie and Heathcott, *Beyond the Ruins*, 237–56; High and Lewis, *Corporate Wasteland*, 77–78.

41. Iron Range Enterprises, *The Mesabi-Vermilion Iron Range*, 11; MGL & Associates, *Tower-Soudan State Park: Interpreting the Soudan Underground Mine* (Minneapolis: MGL & Associates, 1982), 6–13.

42. MGL & Associates, *Tower-Soudan State Park*, 1, 5; Michael Eliseuson, *Tower Soudan: The State Park Down Under* (St. Paul: Minnesota Parks Foundation, 1976), 23–30.

43. IRRRB Meeting Minutes, May 11, 1979, 1–33; June 16, 1979, 6; "Calumet: Hill Annex Mine Added to Program," *Duluth News-Tribune*, February 24, 1979.

44. Architects IV Fugelso, Porter, Simich, Whiteman, *Restoration Plan for Old Calumet, Minnesota* (Duluth, Minn.: Architects IV Fugelso, Porter, Simich, Whiteman, 1974), 1; Community Histories: Calumet, Minnesota, Collection; Iron Range Research Center, Chisholm, Minnesota. On a very similar project to transform a deindustrializing mining community in Ontario into a historical tourist destination, see Pamela Stern and Peter V. Hall, "Historical Limits: Narrowing Possibilities in 'Ontario's Most Historic Town,'" *The Canadian Geographer/Le Géographe Canadien* 54, no. 2 (2010): 209–27.

45. "Calumet," undated manuscript, 2, in Community Histories: Calumet, Minnesota, Collection; Iron Range Research Center, Chisholm, Minnesota; Robert Scott, quoted in IRRRB Meeting Minutes, April 10, 1973, 42–43.

46. Architects IV, *Restoration Plan for Old Calumet, Minnesota*, 1.

47. Ibid., 11, 13, 20, 23, 25, 30, 30–33, 34.

48. KDAL Television 3/610 Radio, "A Statement of Editorial Opinion," September 15, 1971; Community Histories: Calumet, Minnesota, Collection; Iron Range Research Center, Chisholm, Minnesota; Architects IV, *Restoration Plan for Old Calumet, Minnesota*, 34.

49. "Calumet," undated manuscript, 3–4, 8. It is unclear if planners researched whether Calumet had phone booths and antennas in the 1920s.

50. Ibid., 4.

51. Architects IV, *Restoration Plan for Old Calumet, Minnesota*, 37–38.

52. On the 1980s steel crisis, see John P. Hoerr, *And the Wolf Finally Came: The Decline of the American Steel Industry* (Pittsburgh: University of Pittsburgh Press, 1988); William T. Hogan, *World Steel in the 1980s: A Case of Survival* (Lexington, Mass.: Lexington Books, 1983); Sherry Lee Linkon and John Russo, *Steeltown U.S.A.: Work and Memory in Youngstown* (Lawrence: University Press of Kansas, 2002); Jack Metzgar, *Striking Steel: Solidarity Remembered* (Philadelphia: Temple University Press, 2000); and Stein, *Running Steel, Running America*.

53. Steven Greenhouse, "An Ore Carrier's Troubled Odyssey," *New York*

Times, July 14, 1985. On the global iron-ore market up to 1971, see Manners, *The Changing World Market for Iron Ore, 1950–1980.*

54. William E. Schmidt, "Heavy Industry Is Up, and So Is a Corner of Minnesota," *New York Times,* November 20, 1988; William Serrin, "Recession Silences the Mines of the Minnesota Iron Range," *New York Times,* August 7, 1982; Babbitt Embarrass Area Development Association, *Babbitt, Minnesota, Fact Book* (Babbitt, Minn.: Babbitt Embarrass Area Development Association, 1984), 22, in Community Files, Babbitt, Minnesota, Iron Range Research Center, Chisholm, Minnesota; Bill Cook, oral history interview.

55. Schmidt, "Heavy Industry Is Up, and So Is a Corner of Minnesota"; Brown, *Overburden,* 12.

56. Iver Peterson, "Homeless Crisscross U.S. until Their Cars and Their Dreams Break Down," *New York Times,* December 15, 1982.

57. Marsha Benolken, oral history interview, Nashwauk, Minnesota, August 9, 1986, Unemployment on the Iron Range Oral History Collection, A-86–692, Iron Range Research Center, Chisholm, Minnesota.

58. Quoted in D. J. Tice, "The Thing on the Hill," *Corporate Report Minnesota,* October 1982, 66.

59. Bill Cook, oral history interview.

60. Dana Miller, "Public Policy and Economic Development: The Iron Range Experience," paper presented at the Minnesota Historical Society Annual Meeting (Minneapolis, 1990); Betty Wilson, *Rudy! The People's Governor: The Life and Times of Rudy Perpich* (Minneapolis: Nodin, 2005); Paul Delaney, "Rudy Perpich: From the 'Dumps' to Governor," *New York Times,* December 7, 1976; Dewar, "Development Analysis Confronts Politics," 291–92.

61. IRRRB, *Biennial Report, 1982–1984* (St. Paul: IRRRB, 1984); IRRRB, *Biennial Report, 1984–1986* (St. Paul: IRRRB, 1986), 7; a copy of this advertisement is available in the Ironworld hanging folder, Iron Range Research Center, Chisholm, Minnesota.

62. Matthew Frye Jacobson, *Roots Too: White Ethnic Revival in Post-Civil Rights America* (Cambridge: Harvard University Press, 2006); IRRRB, *Biennial Report, 1980–1982,* 11.

63. Nemanic, *One Day for Democracy.*

64. IRRRB, *Biennial Report, 1978–1980,* 6, 8; IRRRB, *Biennial Report, 1980–1982,* 13.

65. Goin and Raymond, "Recycled Landscapes," 271.

66. Midwest Research Institute, *Tourism Development Components of the Iron Range Interpretative Program,* 34–36, 20.

67. Carol R. Sheppard, "Interpretative Center Preserves Iron Range Heritage," *Mining Congress Journal* (September 1980): 27.

68. IRRRB, *Biennial Report, 1980–1982,* 21; J. T. Gehrke, "Ironworld Pursuing Corporate Sponsorship," *Hibbing Daily Tribune,* May 7, 1998; IRRRB, *1993–1994 Biennial Report* (Eveleth, Minn.: IRRRB, 1994), 4, 24.

69. IRRRB, *Biennial Report, 2005–2006* (Eveleth, Minn.: IRRRB, 2006), 13.

70. Janna Goerdt, "Giants Ridge," *Mesabi Daily News,* August 23, 2002. On the business structure of north-woods tourism, see Shapiro, *The Lure of the North Woods.*

71. "Iron Range Solution: Maybe It's with Tourists," *Chicago Tribune*, May 28, 1985; "Mining Nature's Bounty," *Parks and Recreation*, February 2005; IRRRB, *Biennial Report, 1982–1984*.

72. Greg Wong, "Giants Ridge Beautiful, Playable," *St. Paul Pioneer Press*, September 3, 1997.

73. Nemanic, *One Day for Democracy*, 157–58; Mike Binkley, "Finding Minnesota: Alpine Village on the Iron Range," *CBS Minnesota*, December 1, 2013, http://minnesota.cbslocal.com/2013/12/01/finding-minnesota-alpine -village-on-the-iron-range/; Phil Davies, *Scenic Driving Minnesota* (Helena, Mont.: Falcon Publishing, 1997), 117.

74. Bob Geiger, "Town of Biwabik Drawing Attention of Developers," *Finance and Commerce*, June 16, 2005; Hugh E. Bishop, "Riding the Roller Coaster," *Lake Superior Magazine*, March 2001, 20.

75. Binkley, "Finding Minnesota"; Jennifer Bjorhus, "State-Subsidized Giants Ridge Ski Resort Leaves Trail of Red Ink," *Minneapolis Star Tribune*, April 5, 2015; Geiger, "Town of Biwabik Drawing Attention of Developers."

76. On the history of the Keweenaw Peninsula, see Lankton, *Cradle to Grave*; Lankton, *Beyond the Boundaries*; and Lankton. *Hollowed Ground*; Bode Morin, *The Legacy of American Copper Smelting: Industrial Heritage versus Environmental Policy* (Knoxville: University of Tennessee Press, 2013), 163–66.

77. "Calumet May Have Last Laugh," *Daily Mining Gazette* (Houghton, Michigan), February 11, 1987; "Legislation for Park Introduced by Davis," *Daily Mining Gazette*, August 3, 1991; William O. Fink, "Keweenaw National Historical Park Update," August 1993, in National and State Parks 1990–1994 Folder, Box 20, Copper Country League of Women Voters Collection, MS-024, Michigan Tech Archives, Houghton, Michigan.

78. Morin, *The Legacy of American Copper Smelting*, 165; Karen Grassmuck, "Park Leader Backing Keweenaw," *Daily Mining Gazette*, September 13, 1994. Fink, "Keweenaw National Historical Park Update"; Patrick Jasperse, "Parks Are Prime for Pork-Barrel Politics," *Milwaukee Journal Sentinel*, June 23, 1996.

79. Roger Komula, "President Signs Park Bill," *Daily Mining Gazette*, October 28, 1992; R. V. Langseth, "Dreams Don't Die," *Daily Mining Gazette*, October 7, 1992; C. Holleyman, letter to the editor, *Daily Mining Gazette*, May 18, 1993.

80. Pamela Porter, "House Panel Votes to Cut Park Funds," *Daily Mining Gazette*, April 30, 1992; Nick Healy, "Sen. Levin Predicting Park Will Get Funding," *Daily Mining Gazette*, August 23, 1993; Karen Grassmuck, "Funding Fight," *Daily Mining Gazette*, October 27, 1993; Fink, "Keweenaw National Historical Park Update"; Timothy Noah, "Tired of Mountains and Trees? New Park Features Superfund Site, Shopping Mall," *Wall Street Journal*, July 28, 1995. Note that park director Bill Fink responded to Timothy Noah's story in the local newspapers by noting that Noah spent just one day touring the park, seemed interested and excited while visiting, and then returned home to write a scalding attack (Brady Walters, "Official Criticizes Bleak Article on Park," *Daily Mining Gazette*, August 12, 1995).

81. Dante Chinni, "An Uncertain Future in Iron Country," *Christian Science Monitor*, November 21, 2000.

82. Cowie, *Capital Moves*, 182; Steven High, "Capital and Community Reconsidered: The Politics and Meaning of Deindustrialization," *Labour/ Le Travail* 55 (2005): 186.

83. Barry Bluestone and Bennett Harrison, *The Deindustrialization of America: Plant Closings, Community Abandonment, and the Dismantling of Basic Industry* (New York: Basic Books, 1982). For a review of Bluestone and Harrison's role in fomenting broad public concern with deindustrialization in the early 1980s, see Jefferson Cowie and Joseph Heathcott, "Introduction: The Meanings of Deindustrialization," in Cowie and Heathcott, *Beyond the Ruins*, 3–4; Steven High, "Capital and Community Reconsidered;" D. J. Tice, "The Thing on the Hill," 61.

CONCLUSION

1. "Company News: LTV to Close Mining Operation in Minnesota," *New York Times*, May 25, 2000; Tom Scheck, "Mine Closing Rocks Iron Range," Minnesota Public Radio, May 24, 2000, http://news.minnesota.publicradio .org/features/200005/24_scheckt_mine/; David Phelps and Larry Oakes, "Upkeep Is Mine's Downfall," *Minneapolis Star Tribune*, May 25, 2000; Alison Grant, "Bad Luck on Mesabi Range," *Cleveland Plain Dealer*, March 18, 2001, http://www.cleveland.com/indepth/steel/index.ssf?/indepth/steel/more/ fs18mine.html.

2. Amanda Paulson, "Surprise Revival for Iron Mines of Minnesota," *Christian Science Monitor*, April 22, 2004, LexisNexis Academic; Robert Whereatt, "Ventura Offers Miners Help, Hope," *Minneapolis Star Tribune*, June 7, 2000, Proquest Newsstand.

3. Mike Meyers, "New Details Emerge on EVTAC Revival Plan," *Minneapolis Star Tribune*, November 27, 2003, Proquest Newsstand; Jonathan Katz, "Where Does Steel Go Now?" *Business Week*, February 2008, 43; Alice Cantwell, "Court OKs EVTAC Mining Sale," *Daily Deal* (New York), November 26, 2003, www.lexisnexis.com; Paulson, "Surprise Revival for Iron Mines of Minnesota."

4. "Minnesota's Iron Age," *Mining Magazine*, April 2007, 14; Paulson, "Surprise Revival for Iron Mines"; IRRRB, *Biennial Report, 2005–2006* (Eveleth, Minn.: IRRRB, 2006), 5.

5. Larry Oakes, "A Renaissance on the Iron Range," *Minneapolis Star Tribune*, March 1, 2008; "$300 Million Expansion Planned for Keetac Taconite Operation," *Duluth News Tribune*, February 1, 2008.

6. Catherine Conlan, "Sudden, Painful Iron Range Slump Evokes Talk of the '80s," *Minnpost.com*, March 5, 2009, http://www.minnpost.com/stories/ 2009/03/05/7143/sudden_painful_iron_range_slump_evokes_talk_of_the _80s.

7. Josephine Marcotty, "Minnesota's Mining Boom: New Riches or New Threat?" *Minneapolis Star Tribune*, September 25, 2011, LexisNexis Academic; Jim Miller, "Geology and Mineral Deposits of the Duluth Complex, Minnesota and Why It Will Be Mined Someday," Presentation at the Understanding the Impacts of Mining in the Western Lake Superior Region Work-

shop, September 12, 2011, http://mn.water.usgs.gov/projects/tesnar/2011/Presentations/MillerDC%20Min_USGS%20workshop.pdf; Rob Delaney and Stewart Bailey, "Abandoned North American Mines Getting a Second Look," *Toronto Globe and Mail*, August 20, 2008, LexisNexis Academic.

8. Josephine Marcotty, "Battle Waged over Mining Firms' Plans," *Minneapolis Star Tribune*, May 23, 2012, LexisNexis Academic; Marcotty, "Minnesota's Mining Boom"; Josephine Marcotty, "Up North, a Last-Ditch Pitch to Stop Copper Leases," *Minneapolis Star Tribune*, October 5, 2011, LexisNexis Academic.

9. "Morse Town Hall Overflows for County Mining Resolution Debate," *Ely Echo*, December 24, 2011, LexisNexis Academic; Josephine Marcotty, "Clash over Mining Grips North Woods," *Minneapolis Star Tribune*, June 16, 2013, LexisNexis Academic.

10. Lee Bloomquist, "PolyMet Plans New Mine near Hoyt Lakes," *Duluth News-Tribune*, September 15, 2005, LexisNexis Academic; Marcotty, "Minnesota's Mining Boom."

11. Marcotty, "Clash over Mining Grips North Woods."

12. Allison Sherry, "Hot Issues, Uneven Growth Have 8th District Voters Restless," *Minneapolis Star Tribune*, June 7, 2014, http://www.startribune.com/politics/statelocal/262254441.html; Briana Bierschbach, "Capitalizing on Polymet Rift, Republicans Mining DFL Votes on Iron Range," *Minnpost.com*, June 13, 2014, http://www.minnpost.com/politics-policy/2014/06/capitalizing-polymet-rift-republicans-mining-dfl-votes-iron-range; Aaron J. Brown, "GOP's Zellers: 'We Are All Iron Rangers,'" *Minnpost.com*, February 26, 2014, http://www.minnpost.com/minnesota-blog-cabin/2014/02/gop-s-zellers-we-are-all-iron-rangers.

13. Bastow, *"This Vast Pollution . . . ,"* 104–6.

265

cut by, 83, 218; as local issue, 214; lost wages due to, 221; as national problem, 144, 214; scholarship of, xxvii–xxviii, 219, 232n18, 233nn26–27; technology and, xx, 219; unions and, xxiii–xxiv, 125; in United States, xxviii–xxix. *See also* decline

Democratic-Farmer-Labor Party (DFL): copper-nickel mining conflict, 226–27; deindustrialization's effect on, 218; industrial workers in, 56, 66–67, 82; political strength on Iron Range, 56; taconite amendment conflict, 34, 57–61, 63–64, 66, 70, 72, 76–77; taxes and, xxii, 38; unions and, xxiii; *United States of America v. Reserve Mining Company* and, 116. *See also* liberalism; radicalism; *United States of America v. Reserve Mining Company*

Department of Conservation, Minnesota, 90–91, 101

Department of Economic Development, Minnesota, 149, 176, 181

Department of Iron Range Resources and Rehabilitation. *See* Iron Range Resources and Rehabilitation Board

Department of Justice, U.S., 106–7, 109

Department of Natural Resources, Minnesota, 116, 173. *See also* Department of Conservation, Minnesota

Department of the Interior, U.S., 100–102

depletion: of copper, 16–17; foreign ore as solution to, 50; general fears of, 15–17; of iron ore, xii, 13–22; job loss from, xxvi

depressed areas, 140, 144–47, 161

Detroit, x, 135, 150, 215

Dewey, Scott, 123

DFL. *See* Democratic-Farmer-Labor Party

distressed areas. *See* depressed areas

Douglas, Paul, 144–45

Douglas J. Johnson Economic Protection Trust Fund, 158. *See also* Iron Range Resources and Rehabilitation Board

drilling, 21, 142

Dublin, Thomas, xxviii, 68, 160

Dudley, Kathryn, 166, 169

Duluth, 53, 73; cancer scare in, 110–11, 126; Iron Range economy's connection to, xiv; Mesabi Iron Company in, 11; National Water Quality Laboratory in, 99; Jeno Paulucci and, 143; residents' pollution concerns, 127; tension with Silver Bay, 120; water supply of, 109–11, 120

Duluth Area Chamber of Commerce, 107

Duluth Complex, 223–25

Duluth Hotel, 103–4

Dylan, Bob, ix–x, 48–49, 231n1

economic development: anticommunism and, 145–46; ARA and national approach to, 147–48; Blatnik's focus on, 94–96; business-like approach to, 158–59; criticism of, 141, 162, 205; environment and, 95–96, 131; entrepreneurship and, 159; federal-level policies, xxiii, 161, 210, 213; higher education as, 152; history and heritage as, 164, 177, 186–87, 194, 196, 203, 205–6, 211, 215; local focus of, xxiii, 135–36, 148, 160–61; 1960s taconite boom and, 148; possibilities for, in 1950s, 142; refusal to accept decline of, 136; as response to deindustrialization, xx–xxiii, 218–19; state-level, 137; steel crisis and, 202–3; tourism as, 179, 207, 213

Economic Development Administration (EDA), 147, 181, 207

Fraser, Donald, 75, 77
Freeman, Orville, 25
Frick, Henry C., xiv
Fride, Edward T., 103
Furness, Edward M., 97, 104

Gary, Elbert H., xiv
Gary (Indiana), 85, 215
General Electric, 3, 9
ghost towns, xii, 19, 136, 217
Giants Ridge, 206–9
Gilbert (Minnesota), 48, 153
globalization: foreign ore as early
    example of, 50–51; during late
    twentieth century, 79; responses
    to, 154, 218–19; during steel crisis
    of 1980s, 198–99; during twenty-
    first century, 220–23
Goin, Peter, 168, 205
gold mining, xiii, 6
Graham, Otis, Jr. 135, 160
Great Depression, 95–96; IRRRB's
    origins during, 136, 139, 140, 148;
    mine shutdowns during, xv, 19;
    taxation during, 41–42, 55
Great Society, 134, 147
green water, 104–5, 130. *See also*
    pollution; tailings; *United States
    of America v. Reserve Mining
    Company*

Hanna Mining Company, 50–51,
    78, 201
Hays, Samuel, 92, 119
Heathcott, Joseph, xi
hematite: decline of mining of, xix,
    32, 45–47, 81–82, 133, 219; defini-
    tion of, 1–2, 5–6; depletion of, xix,
    13–22, 183; foreign, 45–46, 49–51;
    taxation of, 49
heritage: as apolitical, 164, 166, 216;
    architectural, 166, 175, 210; con-
    flicted, 169; deindustrialization
    and, 214–15, 219–20; diversify-
    ing economy, 164; ethnicity
    and, 204–5; federal support for,
    209–13; industrial preservation

of, 168, 175, 210; in industrial
    regions, xxv, 163, 165, 213–14; in
    Lowell, Massachusetts, 165–66;
    movement in twentieth cen-
    tury, 165–69; during 1980s steel
    crisis, 202; tensions over, 163–64,
    166–68, 175; tourism and, xxv,
    177–79, 211. *See also* memory;
    museums; Old Calumet Restora-
    tion Project; tourism
Hibbing, x, 19, 61, 65, 70, 178; Carey
    Lake, 172–73, 257n15; churches,
    48; Bob Dylan's home, ix, 48–49;
    high school, 42, 47; industrial
    park, 154–55; John F. Kennedy in,
    56, 147; mines in, xv; municipal
    spending, 42–44; in 1960s, 47, 73;
    in 1970s, 149; in 1980s, xxiii, 199,
    201; relocation of, 41. *See also*
    Lakewood Industries; Power,
    Victor
Hickok, Lorena, 42, 136
High, Steven, 161, 169, 175, 214–15
Hill, James J., xiv
Hill Annex Mine State Park, 193
Horn, Charles L., 141
House of Representatives, Minne-
    sota. *See* Legislature, Minnesota
House of Representatives, U.S.,
    Committee on Public Works, 94,
    98–99. *See also* Blatnik, John
Hovis, Logan, 6
Hoyt Lakes (Minnesota), 31, 220,
    225–26
Huffman, Thomas, 89
Hull-Rust-Mahoning Mine, xvi
Humphrey, Hubert H., 77, 97, 108,
    143

Illinois, 144, 147
immigration, xiv–xv; European, xii,
    xvii; Iron Range culture shaped
    by, xvii, 232n14; labor movement
    and, 36
industrial policy, xvii, 234n29; in
    Europe, 134, 144; lack of, 134;
    local- and state-level, 135

industrial sublime, 15

industrial workers: cultural marginalization of, 169; politics of, 82–83; role within DFL, 56, 66–67; social conservatism of, 83; taxation, 37, 82–83

industry: centrality in U.S. history, 231nn4–5; environmental costs of, 87, 132; government and, xvii; inevitability of decline in, xi, 228–29; permanence of, xi, 228–29; resource-extractive, xiii, xxi, xxvi. *See also* deindustrialization; steel industry

Ironman Memorial, 190. *See also* statues

iron ore: changes in sources of, 3, 30, 32, 45; discovery of, 21; foreign, xi, 7, 20, 49–51, 55; geology of, xii–xiii, 1–2, 5–6; global market for, 45–46, 79; under Hibbing, 41; imports of, 157; investment in pelletizing of, 78–79; market changes during *United States of America v. Reserve Mining Company*, 113; in New Jersey, 7; in New York, 7, 9; steel crisis's effect on, 198–99; taxation of, 22–26. *See also* hematite; taconite

Iron Range: Area Redevelopment Administration and, 147–48; boundaries of, xviii; culture of, xvii–xviii, 19; as declining region, 135, 140, 217, 227; as depressed region, 140, 147; desire to sustain, 136; diaspora, xvii, 140; diversification of economy, 139, 142, 148, 164; early history of, xiii–xv; entrepreneurial spirit on, 159–60; fiscal problems of, 49; geology of, xii–xiii; during Great Depression, 137; during Great Recession, 222–23; identity, xvii–xviii; junk on, 150; John F. Kennedy's connection to, 146–47; versus Keweenaw Peninsula, 210; landscape of, 172–73, 175;

as liberal, 56, 83, 217–18, 226–27; marginalization of, xi; municipal spending on, 138; versus other industrial regions, x, xviii, 218, 227–28, 232n15; physical appearance of, 152; political apathy on, 227; as productive industrial region, 164, 173–75; residents' attitudes, 21, 23–24; sense of decline in 1950s, 47–49; significance of, 188, 222; structural shift in 1980s, 157, 202; in twenty-first century, 220–23, 227; voting bloc, 140. *See also* Mesabi Range

Iron Range Interpretative Center. *See* Ironworld

Iron Range League of Municipalities, 61

Iron Range Research Center (IRRC), 186–88

Iron Range residents: ambivalence about taconite amendment, 73–74; anti-environmental attitudes, 100, 122, 226; attitudes toward government, 158–59, 218; attitudes toward taxation, 40–41, 43–45; change in attitudes due to 1980s steel crisis, 202; hostility toward tourism, 173–75; quality of life in 1950s, 47–49; security during taconite boom, 156

Iron Range Resources. *See* Iron Range Resources and Rehabilitation Board

Iron Range Resources and Rehabilitation Board (IRRRB), xxii–xxiii, xxix; ARA and, xxii, 144–48; audit of, 153; beautification programs, 150; continuing New Deal planning tradition, 141; corporate ethos of, 157–59; critics of, 139, 148–49, 153, 155–56, 158; diversification of economy, 139, 142, 148, 156–57; economic development policy of, 135–36, 218; economic planning example, 160–61; entrepreneur-

platinum, 223

pollution: air, 114; employment in abatement of, 121; enforcement conferences to regulate, 97, 102–6; engineers' attitudes toward, 128; history in mining, 89; laws regulating, 92–93; local versus federal regulation of, 99; regulation of, as political pork, 98–99; Reserve Mining trial as key example of, 86–87; sulfides, 224, 226, 228; taconite and, xix–xx, 223; unions protesting, 123–24. *See also* tailings; *United States of America v. Reserve Mining Company*

Pollution Control Agency, Minnesota, 90, 101, 116, 128

Polymet Corporation, 225–26

population, 227; decline in 1950s, 47–48; decline in 1980s, 199–200; increases due to taconite, 149

populism: liberalism and, 69; municipal spending as, 42, 44; Victor Power exemplifying, 41

poverty, 134, 146

Power, Victor, 41, 43–44, 80, 82

presidential campaign of 1960, 145–46

Progressive Era, 40, 134

property taxes, 40–41

public history, xxv, 165–66. *See also* heritage; memory; museums

Public Works, U.S. House of Representatives Committee on, 94, 98–99. *See also* Blatnik, John

Purdue University, 3, 31

Quie, Al, 154

radicalism: Blatnik espousing, 96; deindustrialization and, 161; pragmatism behind, 51–52; rejection of, 74, 159; taconite amendment as reversal of, 60–61, 73, 83; taxation and, 39–40

Raymond, C. Elizabeth, 168, 205

Reagan administration, 125

regional planning, 134. *See also* Iron Range Resources and Rehabilitation Board

Report of the National Conservation Commission (1909), 16

Republican Party: attacking DFL on taconite amendment, 71; in Congress, 212; copper-nickel mining and, 226–27; support for taconite amendment within, 38, 56–57, 67; tax debate (1961), 53

Republic Steel, 23, 26, 107, 113–14

Reserve Mining Company, 31, 52, 61, 223; asbestos claims against, 109–11; assuring workers about pollution, 117–18; bankruptcy of, 132; Blatnik's support for, 96–97; conservationist record of, 91–92; construction of, 27–28, 45; deep pipe plan, 104–6, 112–13; defense strategy of, 103–4; as economic development, 131; enforcement conferences for pollution, 102–6; environmental permits for, 90–91, 100, 224; E. W. Davis Works and, x, xx, 1–3, 28–30, 85–86, 228; executives' defense of, 118–19; formation of, 26–27; Miles Lord's order to close, 114; on-land tailings disposal, 104–6, 112–13, 115–17; production by, 30, 32; shutdown's consequences, 120–21; Silver Bay as company town, 117; support for taconite amendment, 64; Stoddard Report on, 100–102. *See also* Davis, Edward W.; *United States of America v. Reserve Mining Company*

resorts, 170–72, 174. *See also* Giants Ridge

resource-extraction industries: forestry, xiii, 142; poverty and, 146; rural/urban divide over, 121–22; taxation of, xxi, 40; unsustainability of, 136, 217

restaurants, 159–60
revitalization, 165
Rivers and Harbors Act (1899), 100
Rivers and Harbors Subcommittee, 98
Roach, Gene, 127
Rockefeller, John D., xiv
Rolvaag, Karl, 77
Roosevelt, Eleanor, 42, 136
Roosevelt, Franklin, 134, 137
Rosner, David, 123
Ross, James, 24
Rotella, Carlo, 164–65, 167–68
Rukavina, Tom, 226
rural issues, 42, 95, 121–22, 126, 159, 252n3
Rust Belt, 150, 168, 199, 214

Safe Drinking Water Act, 93
Saturday Evening Post, 170
Savage, Kirk, 191
Save Lake Superior Association, 107
School of Mines, University of Minnesota, xix, 3, 8–9
schools. *See* education
Scott, Robert, 185
Segal, George, 191
Selikoff, Irving, 110–11
Senate Committee on Labor and Public Welfare, 145
Serra, Richard, 191
service sector, xxiv–xxv, 164; low wages in, 174–75; training for jobs in, 171–72
Shapiro, Aaron, 171
Sierra Club, 106
silica, 12–13
Silver Bay (Minnesota), 108; as company town, 117–18; construction of, 28, 97; Edward W. Davis's home in, 28, 52, 127–28; E. W. Davis Works in, xx, 2, 3, 30, 85–86, 88–90, 92; harbor at, 91, 100; before Reserve Mining, 27; residents' attitudes, 116; response to cancer scare, 110–11; response to closure of E. W. Davis

Works, 119, 132; response to pollution trial, 101, 117, 119–20, 122, 125–28; Rocky Taconite statue, 75, 189; tailings near, 89, 104, 112–17; tension with Duluth, 120. *See also* E. W. Davis Works; Reserve Mining Company; *United States of America v. Reserve Mining Company*
Skalko, Gary, 223
skiing, 170, 174, 206–8
Smith, Duane, 128
snowmobiling, 170
Social Security Act of 1935, 134
Society for Industrial Archeology, 168
Sofchalk, Donald, 37
Soudan Underground Mine, 191–93
South America, 20, 50, 157, 198–99
Stanton, Cathy, 166
Stassen, Harold, 137
State Senate, Minnesota. *See* Legislature, Minnesota
statues, 191; *The Emergence of Man through Steel,* 184, 188, 190–91; Rocky Taconite, 75, 189
steel crisis (1980s), 120, 132, 157; effect on individuals, 201–2; effect on Iron Range, 199–202; Iron Range's identity changing during, 202; IRRRB's response to, 154, 156, 206–9; national response to, 215; reasons for, 198; unemployment due to, 199. *See also* steel industry
steel industry: Bessemer process, 12–13; depressed areas legislation and, 145; Duluth steel mill, 121; environmental movement and, 118–19; financing taconite mills, 55–56; globalization of, xi, 50–51, 221–22; Iron Range's importance to, 137, 222; during nineteenth century, xiv; supply and demand of, xxvi; taconite investment during 1960s, 71–72, 78–79; taxation of, 23, 25; during

during 1980s steel crisis, 154,
156, 199–202; steel companies
taking advantage of, 58–59;
structural, 148–49, 159; after
taconite amendment, 81, 148–49;
training programs for, 145; used
to lure new development, 151. *See
also* employment
unions: antipollution measures of,
126; divide over taconite amendment, 61–62, 64, 68, 72–73;
environmentalism and, 122–27;
high pay due to, 201; history on
Iron Range, 36–37; response to
deindustrialization, 68; response
to *United States of America v.
Reserve Mining Company,* 122,
127; during World War II, xvi, 37.
See also labor liberalism; United
Steelworkers of America
United Auto Workers (UAW), 123
United Mine Workers of America
(UMW), 68, 124
*United States of America v. Reserve
Mining Company,* xx; appeal to
Eighth Circuit Court of Appeals,
114–16; asbestos controversy
during, 108–11; beginning of,
106–8; Blatnik harmed by, 95;
context of 1970s taconite boom,
120; as corporate coverup, 113;
decline as context for, 87–88;
effect on steel industry, 119;
end of, 130–31; microcosm of
carcinogenic landscape, 110;
order to close Reserve Mining,
114; precedents from, 106; role
in environmental movement,
86–87; rural/urban split due to,
122; significance of, 106, 132–33;
unions' response to, 122, 125–27
United Steelworkers of America
(USWA), xxiii, 213; environmentalism and, 123–24; history on
Iron Range, 36–37; Ironworld
and, 184; organization of, xvi;
rank-and-file's attitude toward

pollution, 127; response to
*United States of America v. Re-
serve Mining Company,* 125, 127;
during taconite amendment, 58,
64, 68–73
United Taconite, 221
University of Minnesota, ix, 11, 23,
28, 31, 52, 76, 107, 131
University of Minnesota Duluth, 152
Upper Great Lakes Regional Commission, 181
urban renewal, 160
U.S. Steel Corporation, 26; creation of, xiv; Duluth mill, 121;
finances of, 55; Keetac, 222–23;
layoffs in 1950s, 47; Minntac,
71, 78; Richard Serra sculpture,
191; Soudan mine, 191; taconite
amendment, 58–60, 66, 69–72;
in Venezuela, 50. *See also* Oliver
Mining Company
USWA. *See* United Steelworkers of
America
Utah, 13, 199

Venezuela, 50
Ventura, Jesse, 220–21
Verity, C. William, 113–14, 118. *See
also United States of America v.
Reserve Mining Company*
Vermilion Range, xviii, 151, 223, 226;
early history of, xiii–xiv; Tower
Soudan mine on, 191
Vermilion State Junior College, 152
Virginia (Minnesota), 23, 47–48, 80,
149
Voyageurs National Park, 176

Wallace, Mike, 164
Wall Street Journal, 213
Ward, Ian, 154
War on Poverty, 147
War Production Board, 19–20
Washington, D.C., 83
water, 14, 88–95, 97–100, 109–11. *See
also* pollution; Superior, Lake
Watergate, 114

A native of Anoka, Minnesota, Jeffrey T. Manuel received his PhD in history from the University of Minnesota. He is associate professor of history at Southern Illinois University Edwardsville.